Oscillations and Waves

Student Monographs in Physics

Series Editor: Professor Douglas F Brewer
Professor of Experimental Physics, University of Sussex

Other books in the series:

Microcomputers
  D G C Jones

Maxwell's Equations and their Applications
  E G Thomas and A J Meadows

Fourier Transforms in Physics
  D C Champeney

Oscillations and Waves

R Buckley

*Institute of Sound and Vibration Research,
University of Southampton*

Adam Hilger Ltd, Bristol and Boston

© R Buckley 1985

All rights reserved. No part of this publication may be reproduced, stored in a retrieval system or transmitted in any form or by any means, electronic, mechanical, photocopying, recording or otherwise, without the prior permission of the publisher.

British Library Cataloguing in Publication Data
Buckley, R.
    Oscillations and waves.——(Student monographs in physics)
    1. Waves   2. Oscillations
    I. Title   II. Series
    531'.1133   QC157

ISBN 0-85274-793-4

Published by Adam Hilger Ltd
Techno House, Redcliffe Way, Bristol BS1 6NX, England
PO Box 230, Accord, MA 02018, USA

Printed in Great Britain by Page Bros (Norwich) Ltd

Contents

Introduction		vii
Bibliography		viii
1	**Oscillations**	1
	1.1 The Equation of Motion of a Simple Pendulum	1
	1.2 Solution of the Equation of Motion: Linear Treatment	2
	1.3 Solution of the Equation of Motion: Non-linear Theory of the Period	5
	1.4 Damped Oscillations	7
	1.5 The Forced Oscillator and Resonance	12
2	**Normal Modes of Oscillation**	17
	2.1 Coupled Pendulums and Normal Modes of Oscillation	17
	2.2 The Normal Modes of a Stretched String Carrying N Masses	22
	2.3 The Limit $N \to \infty$ and the Continuous String	25
3	**Waves—General Properties and Waves on a Stretched String**	27
	3.1 General Remarks	27
	3.2 The Stretched String	29
	3.3 Waves and Oscillations on a Stretched String	31
	3.4 Energetics	35
4	**Sound Waves**	37
	4.1 The Mechanism of Sound	37
	4.2 The Wave Equation for Sound	37
	4.3 Linearised Sound Waves	39
	4.4 Normal Modes of a Pipe	40
	4.5 Intensity of Sinusoidal Sound Waves	41
	4.6 Reflection of Sound Waves	41
	4.7 The Normal Modes of a Pipe with a Movable End	43

5	**Waves on Water**	**46**
	5.1 Introduction	46
	5.2 Waves on Shallow Water	46
	5.3 Linearised Waves on Shallow Water	48
	5.4 Waves on Deep Water	49
	5.5 Ship Waves	51

Index   **55**

Introduction

This book is an elementary account of the mathematical theory of oscillations and waves, and is intended for first-year students of the physical sciences. In keeping with the aims of this series of monographs, the work is not intended as a comprehensive text. Rather it concentrates on those aspects of the theory that frequently cause difficulty, and includes many examples worked out in detail and fully illustrated.

Chapter 1 describes simple harmonic motion, together with linearly damped and sinusoidally forced oscillators. Normal modes of oscillating systems with two and N degrees of freedom are treated in Chapter 2. The remainder of the book is devoted to waves. Chapters 3, 4 and 5 deal respectively with waves on strings, sound waves and water waves. In Chapter 1 the period of a large amplitude pendulum is calculated, otherwise the treatment is linear.

It is assumed that the reader has some familiarity with calculus, including elementary partial differentiation, and also with complex numbers.

The limited length of the text has necessitated the omission of any discussion of elastic or electromagnetic waves, each of which needs substantial physical introduction. For the same reason, no account is given of waves in two or three spatial dimensions. These and other topics are discussed in the texts by C A Coulson and A Jeffrey and by H J Pain (see Bibliography) and the reader should, having mastered the present material, be capable of tackling these somewhat more advanced texts. At a still more advanced level the texts by G B Whitham and by J Lighthill are to be highly recommended, especially for their treatments of non-linear aspects of the theory.

I would like to thank Maureen Strickland for her excellent typing of the manuscript. The staff of Adam Hilger Ltd, especially Neville Hankins, have been the source of a great deal of helpful advice. My deepest appreciation goes to my wife Anne for her support and encouragement.

Bibliography

Coulson C A and Jeffrey A 1977 *Waves: A Mathematical Approach to the Common Types of Wave Motion* 2nd edn (London: Longman)
Lighthill J 1978 *Waves in Fluids* (Cambridge: Cambridge University Press)
Pain H J 1976 *The Physics of Vibration and Waves* 2nd edn (Chichester: Wiley)
Whitham G B 1974 *Linear and Non-linear Waves* (Chichester: Wiley-Interscience)

Oscillations

1.1 The Equation of Motion of a Simple Pendulum

The elementary theory of the pendulum will be familiar to readers. Here it will be used to introduce the fundamental quantities that occur in oscillatory problems. In addition, its analysis will be carried further. The simple pendulum is a weighted string moving in a vertical plane under the influence of gravity alone. Readers are advised to construct simple models of the systems considered in this chapter and equip themselves with a stop-watch and protractor in order to test approximately the conclusions that are reached.

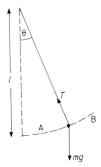

Figure 1.1 The simple pendulum.

Figure 1.1 shows such a pendulum instantaneously displaced through an angle θ (radians) from the vertical. The mass of the bob is m, its weight is mg, where g is the acceleration due to gravity, and its length is l. We are not interested in the tension T in the string, and so consider the resolution of forces perpendicular to the string along the line AB. The appropriate component of the weight, $mg \sin \theta$ attempts to restore the string to its vertical position. The larger θ is, the larger is this component, and it vanishes only when $\theta = 0$. This force will, by Newton's law, cause the mass to accelerate. The velocity of the bob outward along AB is

$$v = l \, d\theta/dt = l\dot\theta \qquad (1.1)$$

where the 'dot' notation is commonly used to denote a time derivative. The acceleration is correspondingly

$$a = l\,d^2\theta/dt^2 = l\ddot{\theta}. \tag{1.2}$$

The quantities $\dot{\theta}$ and $\ddot{\theta}$ are known as angular velocity and acceleration, respectively. Equating force to mass × acceleration, and noting the opposing directions, we find the *equation of motion* of the system

$$ml\ddot{\theta} = -mg\sin\theta \quad \text{or} \quad \ddot{\theta} = -(g/l)\sin\theta. \tag{1.3}$$

The equation of motion having been formulated, it must now be solved to yield θ as a function of time t. Some facts about the equation will, however, be considered first. Since both gravitational and inertial forces (mass × acceleration) are proportional to m, the latter cancels. An immediate prediction is therefore that the motion is independent of the mass of the bob.

Equation (1.3) is a *differential equation* (henceforth DE) of the *second order*. It relates θ to its second time derivative. Equations of motion are usually differential, though most of them are, as we shall see, considerably more complicated than this one. It is vital that the equation be second order. Two arbitrary constants of integration are involved in the general solution of such equations. At some initial time it is possible to release the pendulum from some arbitrary angle θ_0, with some arbitrary velocity $\dot{\theta}_0$. These are called the *initial conditions*. Our equation should be capable of encompassing all such possibilities, and the two arbitrary constants involved in its general solution allow this to be so. Had the equation been of any order other than the second, we could say immediately that the problem had been formulated incorrectly.

1.2 Solution of the Equation of Motion: Linear Treatment

It is a surprising fact that equation (1.3) is in general very difficult to solve. The reason is that $\sin\theta$ is not a *linear* function of θ. We shall outline the exact solution later. For the time being, a process of approximation is introduced that is universally used in the initial stages of such investigations. This is called *linearisation*. A linear differential equation involves the dependent variable (in our case θ) and its derivatives to the first power only. Such equations are very much easier to solve than non-linear ones like equation (1.3).

Are there circumstances in which equation (1.3) can be *approximated* by a linear one? Suppose that the pendulum is allowed to swing through a small angle α. For small θ, $\sin\theta$ is roughly equal to θ. If this is valid, equation (1.3) can be approximated by

$$\ddot{\theta} = -(g/l)\theta \tag{1.4}$$

which is a linear equation.

It is vital to establish limits to the validity of such approximations. To what

Oscillations

extent does sin θ differ from θ when θ is small? We have by Taylor's theorem

$$\sin\theta \simeq \theta - \tfrac{1}{6}\theta^3 \qquad \text{so} \qquad 1 - (\sin\theta/\theta) \simeq \tfrac{1}{6}\theta^2. \tag{1.5}$$

If we require $\sin\theta/\theta$ to be unity to within, say, $P\%$ we must restrict $|\theta|$ to values less than α where

$$\alpha^2/6 = P/100$$

or

$$\alpha = (0.06P)^{1/2} \text{ (radians)} = 15P^{1/2} \text{ (degrees)}. \tag{1.6}$$

If $P = 1$, $\alpha = 15°$, and if $P = 0.1$, $\alpha = 5°$ approximately. These are quite substantial angles. We conclude that there are useful circumstances (e.g. the pendulum of a clock) in which our approximation is of significance. We consider therefore the solution of equation (1.4).

The reader will be aware that equation (1.4) possesses solutions involving sines and cosines. We shall approach this by introducing a formalism, the *complex exponential function* which renders the algebra very compact, and is commonly used in such circumstances. Its properties in outline are as follows. Denoting $\sqrt{-1}$ by the symbol i, it follows by comparison of power series that the following fundamental identity holds

$$e^{ix} = \cos x + i\sin x \qquad \text{so} \qquad \cos x = \text{Re}(e^{ix}) \qquad \sin x = \text{Im}(e^{ix}) \tag{1.7}$$

where Re and Im denote the real and imaginary parts. Let C be a complex number of the form $A + iB$. Then

$$C\,e^{ix} = (A + iB)(\cos x + i\sin x)$$

and so

$$\text{Re}(C\,e^{ix}) = A\cos x - B\sin x \tag{1.8}$$

$$\text{Im}(C\,e^{ix}) = A\sin x + B\cos x.$$

By such means, $\cos x$ and $\sin x$ can be incorporated in the single function e^{ix}.

We attempt to find a solution of equation (1.4) having the form

$$\theta = \text{Re}(C\,e^{i\omega t}) \tag{1.9}$$

where ω and C are (possibly complex) constants. In this case

$$\ddot{\theta} = \text{Re}[(C\,d^2/dt^2)e^{i\omega t}] = \text{Re}(-\omega^2 C\,e^{i\omega t}) \tag{1.10}$$

and equation (1.4) becomes

$$\text{Re}\,[(\omega^2 - g/l)C\,e^{i\omega t}] = 0. \tag{1.11}$$

Now, if equation (1.9) is to be a solution, then equation (1.11) must be identically true for all values of t. Since $C \neq 0$, this will be so only if

$$\omega = \sqrt{g/l}. \tag{1.12}$$

If ω has this real value, equation (1.9) is a solution of equation (1.4) whatever value the complex constant C takes. C is called the *complex amplitude* of the motion, and since it involves two arbitrary real constants, we have in equation (1.9) precisely that feature necessary for a general solution. Thus, if $C = A + iB$, then equations (1.9) and (1.8) show that

$$\theta = A \cos \omega t - B \sin \omega t. \tag{1.13}$$

We can now adjust C so as to satisfy initial conditions. Suppose that when $t = t_0$, we have $\theta = \theta_0$, $\dot{\theta} = \dot{\theta}_0$. Equation (1.9) shows that C must satisfy

$$\theta_0 = \text{Re}(C \, e^{i\omega t_0}) \qquad \dot{\theta}_0 = \text{Re}(Ci\omega \, e^{i\omega t_0}). \tag{1.14}$$

Since $\text{Re}(iZ) = -\text{Im}(Z)$, the latter condition can be written

$$\dot{\theta}_0/\omega = -\text{Im}(C \, e^{i\omega t_0})$$

and hence

$$C \, e^{i\omega t_0} = \theta_0 - i\dot{\theta}_0/\omega \tag{1.15}$$

and the complex constant C is determined. Substituting this value into equation (1.9) yields the solution of the problem

$$\theta = \text{Re}\left[(\theta_0 - i\dot{\theta}_0/\omega) \, e^{i\omega(t - t_0)}\right]$$
$$= \theta_0 \cos \omega(t - t_0) + (\dot{\theta}_0/\omega) \sin \omega(t - t_0). \tag{1.16}$$

Readers are strongly advised to follow these complex manipulations through in detail, to familiarise themselves with the method.

The particular solution (1.16) is of less significance than the key result in equation (1.12), which is now emphasised. For a given pendulum, the motion is periodic, with a frequency ω independent of the detailed motion. θ repeats itself when t increases by $2\pi/\omega$. This is the *period* T of the motion

$$T = 2\pi(l/g)^{1/2} \tag{1.17}$$

which depends only on the length l of the pendulum, and g. T is about 2 s when $l = 1$ m. ω is called the angular frequency (rad s^{-1}). The more common frequency, the *number of oscillations per second*, is

$$f = \omega/2\pi = 1/T. \tag{1.18}$$

This is measured in hertz (Hz).

Some properties of equation (1.16) will now be discussed. The time t appears only in the combination $t - t_0$. This is entirely reasonable. Two motions started identically at different times will proceed identically from those times. Henceforth we can therefore simply allow t_0 to be zero. θ then varies with t as sketched in figure 1.2, which illustrates the behaviour when both θ_0 and $\dot{\theta}_0$ are positive. Any such motion is called *simple harmonic*. The maximum amplitude of the motion can be determined as follows, again using complex notation. Let α and ε be real quantities, such that

$$\theta_0 - i\dot{\theta}_0/\omega = \alpha \, e^{-i\varepsilon} \tag{1.19}$$

so that
$$\alpha = |\theta_0 - i\dot{\theta}_0/\omega| = [\theta_0^2 + (\dot{\theta}_0/\omega)^2]^{1/2}$$
and
$$\varepsilon = -\arg(\theta_0 - i\dot{\theta}_0/\omega) = \tan^{-1}(\dot{\theta}_0/\omega\theta_0). \tag{1.20}$$

Then (with $t_0 = 0$), equation (1.16) becomes
$$\theta = \text{Re}(\alpha \, e^{-i\varepsilon} \, e^{i\omega t}) = \alpha \cos(\omega t - \varepsilon). \tag{1.21}$$

Hence the maximum value of θ, the *amplitude* of the motion, is just α. It occurs when $t = \varepsilon/\omega$, and periodically thereafter, with corresponding values of $-\alpha$ at intermediate times. In order for our linearisation to be valid (equation (1.7)), neither θ_0 nor $\dot{\theta}_0$ must be too large.

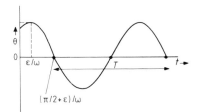

Figure 1.2 Simple harmonic motion: equation (1.21).

The quantity $\omega t - \varepsilon$ is called the *phase* of the motion. Two pendulums oscillating with the same amplitude α but different values of ε_1 and ε_2 will have

$$\theta_1 = \alpha \cos(\omega t - \varepsilon_1) \qquad \theta_2 = \alpha \cos(\omega t - \varepsilon_2). \tag{1.22}$$

If $\varepsilon_1 > \varepsilon_2$, the first follows the motion of the second with a time delay $(\varepsilon_1 - \varepsilon_2)/\omega$. It is said to lag the first in phase by the amount $\varepsilon_1 - \varepsilon_2$; the phase difference between the two. The concepts of frequency, period, amplitude and phase appear in all oscillations and must be thoroughly understood.

This concludes our discussion of the linearised problem. The key conclusions are that the motion is simple harmonic, and that the frequency is independent of the details of the motion.

1.3 Solution of the Equation of Motion: Non-linear Theory of the Period

Experiment will show that for amplitudes of swing greater than a few degrees the period departs from the linearised value $2\pi(l/g)^{1/2}$ increasing with amplitude α as sketched in figure 1.3. It may also be observed that the motion, though still periodic, ceases to be simple harmonic. The bob spends a relatively longer fraction of its swing at larger values of θ.

Both fundamental properties of the linearised solution have therefore to be abandoned. The analysis of the period for a large-amplitude pendulum can be accomplished without too much mathematical difficulty, and will be given here. However, the details of the motion are beyond the scope of this book. Our equation of motion (1.3) is

$$\ddot{\theta} = -\Omega^2 \sin \theta \qquad (1.23)$$

where $\Omega = (g/l)^{1/2}$ is, we recall, the frequency of the linearised system.

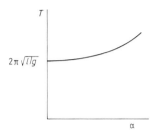

Figure 1.3 Period/amplitude dependence for simple pendulum.

This can be integrated once exactly to yield a first-order differential equation relating θ and $\dot{\theta}$. If we multiply each side by $\dot{\theta}$, and note that

$$d(\tfrac{1}{2}\dot{\theta}^2)/dt = \dot{\theta}\ddot{\theta} \qquad \text{and} \qquad d(\cos \theta)/dt = -\sin \theta \dot{\theta}$$

we can derive the result

$$\tfrac{1}{2}\dot{\theta}^2 = \Omega^2 \cos \theta + \text{constant}. \qquad (1.24)$$

If the amplitude (the maximum value of θ) is α, then $\dot{\theta} = 0$ when $\theta = \alpha$. The constant in equation (1.24) is $-\Omega^2 \cos \alpha$, and we have

$$\tfrac{1}{2}\dot{\theta}^2 = \Omega^2(\cos \theta - \cos \alpha). \qquad (1.25)$$

This equation, besides being a step in the solution of the problem, is of fundamental physical importance in its own right. It can be written in the form

$$\tfrac{1}{2}m(l\dot{\theta})^2 + mgl(1 - \cos \theta) = mgl(1 - \cos \alpha). \qquad (1.26)$$

This is the *law of conservation of energy* for the system. The first term is the *kinetic energy* $\tfrac{1}{2}mV^2$, where $V = l\dot{\theta}$ is the velocity of the bob. The second term is the *potential energy* imparted to the bob in raising it through a vertical distance $l - l \cos \theta$. Equation (1.26) states that the sum of these remains constant throughout the motion. The constant is equal to the kinetic energy when $\theta = 0$, and there is no potential energy, or to the potential energy when $\theta = \alpha$, and there is no kinetic energy.

Returning to equation (1.25), we can express it as an integral

$$\Omega t = \int \frac{d\theta}{[2(\cos \theta - \cos \alpha)]^{1/2}} = \frac{1}{2} \int \frac{d\theta}{(\sin^2 \tfrac{1}{2}\alpha - \sin^2 \tfrac{1}{2}\theta)^{1/2}}. \qquad (1.27)$$

The exact analysis is difficult because, surprisingly, this integral cannot be expressed in terms of familiar functions. When α is small, it is permissible to replace $\sin \frac{1}{2}\alpha$ and $\sin \frac{1}{2}\theta$ by $\frac{1}{2}\alpha$ and $\frac{1}{2}\theta$, and to perform the integral. The linearised trigonometric solution is then eventually recovered. Consider the period T of the motion. It is clear that one quarter of this passes while θ increases from zero to α, during which time $\dot{\theta}$ is positive. Two further quarter periods follow during which $\dot{\theta}$ is negative and θ decreases from α through zero to $-\alpha$. During the final quarter period, $\dot{\theta}$ is again positive and θ increases from $-\alpha$ to zero. T is then given, from equation (1.27) by

$$T = \frac{2}{\Omega} \int_0^\alpha \frac{\mathrm{d}\theta}{(\sin^2 \frac{1}{2}\alpha - \sin^2 \frac{1}{2}\theta)^{1/2}}. \tag{1.28}$$

Formally this expresses T as a function of Ω and α. It is, however, difficult to elicit the dependence on α, because the latter appears both as the upper limit of the integral and inside the integrand. We can manipulate equation (1.28) into a simpler form by introducing a new variable of integration ϕ by means of the substitution

$$\sin \tfrac{1}{2}\theta = \sin \tfrac{1}{2}\alpha \sin \phi. \tag{1.29}$$

While θ increases from 0 to α, ϕ increases from 0 to $\pi/2$. The substitution yields

$$T = \frac{4}{\Omega} \int_0^{\pi/2} \frac{\mathrm{d}\phi}{(1 - \sin^2 \frac{1}{2}\alpha \sin^2 \phi)^{1/2}} \tag{1.30}$$

and α now appears only in the integrand. We must evaluate this approximately, and we do so by assuming that α is not too large and expanding the integrand using the binomial theorem

$$T \simeq \frac{4}{\Omega} \int_0^{\pi/2} \mathrm{d}\phi (1 + \tfrac{1}{2} \sin^2 \tfrac{1}{2}\alpha \sin^2 \phi)$$

$$= (2\pi/\Omega)(1 + \tfrac{1}{4} \sin^2 \tfrac{1}{2}\alpha)$$

$$\simeq 2\pi(l/g)^{1/2}(1 + \tfrac{1}{16}\alpha^2). \tag{1.31}$$

In the last step $\sin^2 \tfrac{1}{2}\alpha$ has been replaced by $\tfrac{1}{4}\alpha^2$, since terms of the next order proportional to α^4 have already been dropped.

Equation (1.31) predicts explicitly the increase of T with α and should be checked against experiment. By expanding equation (1.30) to the *next* order in $\sin^2 \tfrac{1}{2}\alpha$, the reader should be able to evaluate the term in α^4 that is absent from equation (1.31) and thus show that it is accurate to within 1% if α is less than about 50°.

1.4 Damped Oscillations

The amplitude of a freely swinging pendulum decreases, becoming imperceptible after a few tens of periods. This is, of course, due to dissipative forces of *friction*, partly in the pivot and partly on the bob as it moves through the air. All real vibrating systems are so influenced, and the consequences will now be analysed. In this section, an approximate, almost universal, treatment of such effects will be described.

The forces of friction are, on a molecular level, exceedingly intricate, and approximate approaches of one kind or another are necessary. The simplest is that in which it is assumed that the *force of friction opposes and is directly proportional to the velocity of the oscillator*. The resulting equation of motion remains, for small amplitudes, *linear*, and the mathematical analysis straightforward. It should be emphasised however that this is seldom a good approximation for mechanical systems. More realistic forms of damping complicate the analysis enormously, and in any case are usually approached by building on the linear treatment. (It is however possible to construct mechanical oscillators with artificial linear damping imposed. Further, oscillating *electrical circuits*, in which ohmic resistance supplies the damping, are represented realistically in a linear fashion.) For this reason the pendulum in particular will not henceforth be emphasised.

Consider a mass m capable of moving through a distance x, subject to a linear restoring force $-Kx$, and a linear frictional force $-R\dot{x}$, where K and R are positive constants. Its equation of motion will be

$$m\ddot{x} = -Kx - R\dot{x} \tag{1.32}$$

a linear DE of second order. No attempt will be made to estimate appropriate numerical values of R for mechanical systems. Let us introduce the frequency $\Omega = \sqrt{K/m}$ that the oscillator would have in the absence of damping (compare equation (1.12)). Additionally we introduce a quantity $\alpha = R/2m$ to measure the frictional strength. Equation (1.32) then becomes

$$\ddot{x} + 2\alpha\dot{x} + \Omega^2 x = 0. \tag{1.33}$$

The factor 2 simplifies the subsequent algebra somewhat. α has the dimensions of inverse time, and it is clear that the ratio α/Ω, a dimensionless measure of the friction, will play a crucial role.

At this stage we can regard (1.33) as describing *any* oscillator of the present kind. x could represent an angle, or even an electrical voltage, as well as a distance. Whatever the system, the two quantities Ω and α can be defined.

Because the equation remains linear, the apparatus of the complex exponential discussed in §1.2 can be retained (and is indeed now almost essential). Thus a solution of the form

$$x = \mathrm{Re}(C\, e^{i\omega t}) \tag{1.34}$$

is attempted. This satisfies (1.33) for all values of the complex amplitude C, provided that ω satisfies the quadratic equation

$$\omega^2 - 2i\alpha\omega - \Omega^2 = 0. \tag{1.35}$$

This has the two solutions

$$\omega_\pm = i\alpha \pm (\Omega^2 - \alpha^2)^{1/2}. \tag{1.36}$$

(In the absence of the factor 2 in (1.33), inconvenient halves and quarters would appear in (1.36).)

Three cases now must be considered separately according to the relative magnitudes of Ω and α. For reasons which will become apparent these are called *underdamped*, *overdamped* and *critically damped*.

1.4.1 The underdamped case $\alpha < \Omega$

This corresponds to 'small' damping. We introduce the subsidiary quantity v (with dimensions of frequency)

$$v = (\Omega^2 - \alpha^2)^{1/2} \tag{1.37}$$

so that $\omega_\pm = \pm v + i\alpha$. Consider the solution appropriate to ω_+. The exponent $i\omega t$ in (1.34) is

$$i\omega t = -\alpha t + ivt. \tag{1.38}$$

Since αt is real (1.34) becomes

$$x = e^{-\alpha t} \text{Re}(C\, e^{ivt}). \tag{1.39}$$

Note that the *imaginary* part α of ω_+ is responsible for the *real* exponential term $e^{-\alpha t}$. If $C = A + iB$, then

$$x = e^{-\alpha t}(A \cos vt - B \sin vt). \tag{1.40}$$

This contains two arbitrary real constants A and B, and so must be the general solution of (1.33). It is easy to see that, using ω_- instead of ω_+, the same form of solution is obtained. (This is not so for the overdamped case, as we shall see.) In particular, suppose the oscillator starts at $t=0$ with $x=0$ and $\dot{x} = V$. Then $A=0$, $B = -V/v$ and we have

$$x = V e^{-\alpha t} \sin vt/v \tag{1.41}$$

which is sketched in figure 1.4.

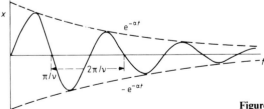

Figure 1.4 Underdamped motion.

In the absence of the term $\exp(-\alpha t)$, this would represent simple harmonic motion (SHM) with frequency v and period $2\pi/v$. v is less than the undamped frequency Ω, which is to be expected; friction will retard the motion. For small damping, with $\alpha \ll \Omega$, we have from (1.37)

$$v \simeq \Omega(1 - \alpha^2/2\Omega^2) \qquad (1.42)$$

and the frequency is reduced only to the second order in α/Ω. Thus if $\alpha = 0.1\Omega$, Ω exceeds v by only 0.5%.

The most significant consequence of damping is however the factor $\exp(-\alpha t)$ which causes the amplitude to decay exponentially to zero. In one period, the motion decays by a factor

$$\exp[-2\pi(\alpha/v)] \qquad (1.43)$$

which, if $\alpha = 0.1\Omega$, is approximately $\exp(-\pi/5) = 0.54$. The motion will decay to 1% of its original value in a time T such that $\exp(-\alpha T) = 0.01$, or

$$T = \ln 100/\alpha \simeq 4.6/\alpha \qquad (1.44)$$

and the number of periods of oscillation in this time is

$$vT/2\pi = (\ln 100/2\pi)(v/\alpha) \simeq 0.73 v/\alpha \qquad (1.45)$$

which is roughly 7 if $\alpha/\Omega = 0.1$. If, on the other hand, say 100 oscillations occur while the motion decays 100-fold, we should have

$$\alpha/\Omega \simeq \alpha/v = 100(2\pi/\ln 100) = 136. \qquad (1.46)$$

The underdamped case occurs then if the resistance is small in the sense that $\alpha < \Omega$. It involves exponentially damped SHM. The rate of damping is proportional to α, and the frequency is decreased proportionally to $(\alpha/\Omega)^2$.

1.4.2 *The overdamped case* $\alpha > \Omega$

This corresponds to 'large' damping, and the solution changes profoundly. The quantity under the square root in (1.36) is now negative, and so this square root is itself imaginary. Introduce in place of (1.37), the subsidiary quantity β

$$\beta = (\alpha^2 - \Omega^2)^{1/2} < \alpha. \qquad (1.47)$$

The two roots $\omega_\pm$ are now both imaginary

$$\omega_\pm = i(\alpha \pm \beta) \qquad (1.48)$$

and the 'complex' exponential $\exp(i\omega t)$ in (1.34) is real in both cases

$$\exp(i\omega_\pm t) = \exp[-(\alpha \pm \beta)t]. \qquad (1.49)$$

Each of the numbers $\alpha \pm \beta$ is positive, so these terms both represent exponential *decay*. The general solution of (1.33) is

$$x = A \exp[-(\alpha - \beta)t] + B \exp[-(\alpha + \beta)t] \qquad (1.50)$$

and retains no trigonometrical (oscillatory) features. Unlike the underdamped case, both ω_+ and ω_- must be considered to obtain the general solution. Suppose again that at $t = 0$, $x = 0$, and $\dot{x} = V$. Then

$$A + B = 0 \qquad -A(\alpha - \beta) - B(\alpha + \beta) = V$$

and so
$$A = -B = V/2\beta. \quad (1.51)$$

Equation (1.50) becomes
$$x = (V/2\beta)\{\exp[-(\alpha-\beta)t] - \exp[-(\alpha+\beta)t]\}$$
$$= (V/\beta)\exp(-\alpha t)\sinh\beta t \quad (1.52)$$

which is sketched in figure 1.5.

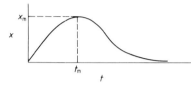

Figure 1.5 Overdamped and critically damped motion.

There are now *no oscillations*. An impulse applied to the system at $t=0$ drives it to a single maximum x_m at $t=t_m$; thereafter the amplitude decays monotonically to zero. Equating $\dot{x}$ to zero, it can be seen that t_m is given by
$$t_m = \frac{1}{2\beta}\ln\left(\frac{\alpha+\beta}{\alpha-\beta}\right). \quad (1.53)$$

An intricate expression results for the corresponding x_m. This, and t_m, are each simplified considerably if a dimensionless measure γ of the strength of resistance is introduced via
$$\alpha = \Omega \cosh\gamma \quad (1.54)$$

whence $\beta = \Omega\sinh\gamma$, from (1.47).

Then t_m and x_m become
$$t_m = \Omega^{-1}\gamma/\sinh\gamma \qquad x_m = V\Omega^{-1}\exp(-\gamma/\tanh\gamma). \quad (1.55)$$

At large enough t, (1.52) is dominated by the first exponential
$$x \simeq V(2\Omega)^{-1}\exp(-e^{-\gamma}\Omega t)/\sinh\gamma. \quad (1.56)$$

The original character of Ω as a *frequency* has now been lost completely. Consider for instance a heavily damped oscillator with $\alpha/\Omega = 3$. Then, from (1.54), $\gamma \simeq 1.7$, and
$$t_m = 0.6/\Omega \qquad x_m = 0.17V/\Omega$$
$$x \simeq 0.18 V\Omega^{-1}\exp(-0.18\Omega t) \quad (1.57)$$

the latter applying to $t \gg t_m$.

1.4.3 The critically damped case $\alpha = \Omega$

The last, and most special, case occurs when $\alpha = \Omega$. In this circumstance it follows from (1.36) that ω_+ and ω_- are each equal to $i\alpha$. Equation (1.34) would then yield

$$x = A e^{-\alpha t}. \qquad (1.58)$$

However, this involves only *one* constant of integration, A, and there must be two such in the general solution. There is a standard mathematical method for dealing with this problem, but it is more illuminating simply to take the appropriate limits of the under and overdamped cases as $\alpha \to \Omega$. Consider the solution (1.41)

$$x = V e^{-\alpha t} \sin vt/v. \qquad (1.59)$$

As α approaches Ω from below, $v \to 0$, $\sin vt \to vt$, and so in the limit (1.59) becomes

$$x = V e^{-\alpha t} t. \qquad (1.60)$$

The same follows from (1.52) as $\alpha \to \Omega$ from above and $\beta \to 0$. Hence a *linear* time factor t appears, which is algebraically quite different from an exponential. The general solution of (1.33), when $\alpha = \Omega$, is

$$x = (A + Bt) e^{-\alpha t} \qquad (1.61)$$

where A and B are constants of integration. The graph of (1.60) is quite similar to (1.52), as shown in figure 1.5. The maximum displacement is $x_m = V/(\alpha e)$ and occurs when $t_m = 1/\alpha$.

In summary therefore, we have considered SHM, subject to linear damping. If the latter is small enough, exponentially damped sinusoidal oscillations result. Above a critical level, however, damping prevents oscillation from occurring and an impulse applied to the system drives it to a single maximum displacement, after which it returns monotonically to zero.

1.5 The Forced Oscillator and Resonance

It is of great practical importance to analyse the response of oscillatory systems to imposed external forces, especially periodic ones. An undamped or lightly damped oscillator possesses, as we have seen, a natural frequency. When subject to a periodic force with frequency widely different from this, one expects that the resulting motion will be small, whereas if this frequency is comparable to the natural one, a large response should occur. This is the phenomenon of *resonance*. It is sometimes exploited, but may also be dangerous, and something to be avoided.

We shall consider again a general linear oscillator subject to a sinusoidal force of given frequency ω. The equation of motion is obtained by appending this force to equation (1.33)

$$\ddot{x} + 2\alpha\dot{x} + \Omega^2 x = F \cos \omega t. \qquad (1.62)$$

We call the constant F the amplitude of the imposed 'force', although its physical dimension will depend on that of x. If, for instance, x is the displacement of a mass m, then this mass will have been absorbed into the number F.

The solution of an equation of such a form is the sum of two distinct parts. The first, known in the theory of differential equations as the 'complementary function', is just the general solution of the equation without the forcing term. This we have already discussed in §1.4. In the presence of damping it decays ultimately to zero as time proceeds. The second, the 'particular integral', is in this case *a sinusoidal oscillation at the forcing frequency* ω. The proportions of these parts present in any particular problem depend on the initial conditions. Whatever these are, however, in the presence of damping the contribution of the first part will eventually become negligible and is called therefore the *transient*. The solution thereafter is dominated by the second part, the *steady state* response, which is independent of the initial conditions. This we shall now analyse by attempting a solution of (1.62) which is sinusoidal with frequency ω

$$x = \operatorname{Re}(X e^{i\omega t}) \tag{1.63}$$

where X is a complex amplitude to be determined. The forcing term in (1.62) can be written $\operatorname{Re}(F e^{i\omega t})$, and it is apparent that (1.63) will satisfy it provided that

$$X(-\omega^2 + 2i\alpha\omega + \Omega^2) = F. \tag{1.64}$$

So the amplitude of response X is determined in terms of the amplitude of forcing, in the frequency dependent fashion

$$X = F/W \quad \text{with} \quad W = -\omega^2 + 2i\alpha\omega + \Omega^2. \tag{1.65}$$

Notice that the *vanishing* of W would be equation (1.35) determining the (complex) natural frequencies of the oscillator. Here, however, ω is real and, in the presence of damping, W never vanishes for such a real value.

Consider a mechanical oscillator of mass m, so that

$$m\ddot{x} + R\dot{x} + Kx = F\cos\omega t. \tag{1.66}$$

The steady state response would be

$$x = \operatorname{Re}(X e^{i\omega t}) \quad \text{where} \quad X = F/(-\omega^2 m + iR\omega + K). \tag{1.67}$$

The *velocity* amplitude is $i\omega X = V$, and

$$V = F/Z(\omega) \quad \text{where} \quad Z(\omega) = i\omega m + R + K/i\omega. \tag{1.68}$$

The quantity Z is called the *mechanical impedance* of the system, and is the ratio of complex amplitudes of applied force and resultant velocity. Its real part is the *resistive* coefficient R, while its imaginary or *reactive* part has a positive term $m\omega$ associated with *inertia*, and a negative term $-K/\omega$ associated with the restoring force or *stiffness* of the system. The electrical analogue for AC circuits is familiar. There the real part of impedance is the ohmic resistance R, while inductance L contributes $i\omega L$, similar to inertia, and capacitance C contributes $1/i\omega C$ akin to stiffness.

Returning to (1.65) we shall analyse the behaviour of the complex quantity W^{-1} as a function of applied frequency ω. Consider the modulus/argument form

$$W = A^{-1} e^{i\phi} \tag{1.69}$$

so that

$$X = FA\,e^{-i\phi} \quad \text{and} \quad x = FA\cos(\omega t - \phi) \quad (1.70)$$

where ϕ is the phase difference between displacement and force and A is the real amplitude ratio between them. Each is a function of ω, and from (1.65) we have

$$\phi = \tan^{-1}[2\alpha\omega/(\Omega^2 - \omega^2)] \qquad A^2 = [(\omega^2 - \Omega^2)^2 + 4\alpha^2\omega^2]^{-1}. \quad (1.71)$$

ϕ is plotted against ω in figure 1.6. It increases from zero when ω is small, through 90° at the undamped natural frequency Ω, to 180° at large ω. The rapidity with which it increases from 0° to 180° can be estimated from those frequencies ω_-, ω_+ at which it is equal, respectively, to 45° and 135° (figure 1.6). Here, $\tan\phi = \pm 1$ and $\Omega^2 - \omega^2 = \pm 2\alpha\omega$. Thus

$$\omega_\pm = (\Omega^2 + \alpha^2)^{1/2} \pm \alpha \simeq \Omega \pm \alpha \qquad \text{if } \alpha \ll \Omega. \quad (1.72)$$

For a lightly damped system, therefore, ϕ increases rapidly from 0° to 180° over a range of frequencies of a few α centred on Ω.

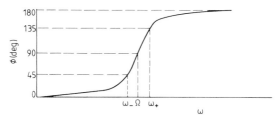

Figure 1.6 Phase difference between displacement and force.

The squared amplitude A^2 is plotted against ω in figure 1.7. Curves 1, 2 and 3 correspond to small, intermediate and large values of the damping ratio α/Ω, in a sense to be considered. A^2 has the value Ω^{-4} at frequencies much less than Ω, and falls to zero as ω^{-4} for $\omega \gg \Omega$. The reasons are clear. At sufficiently small frequencies the response is dominated by the restoring force $-\Omega^2 x$, while at high frequencies it is dominated by the inertia (acceleration) $\ddot{x} = -\omega^2 x$.

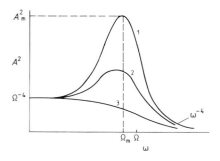

Figure 1.7 Amplitude ratio of displacement and force.

A^2 may have a maximum value at an intermediate frequency Ω_m. It can be located by differentiation, or more straightforwardly by completing the square in (1.71)

$$A^2 = \{[\omega^2 - (\Omega^2 - 2\alpha^2)]^2 + 4\alpha^2(\Omega^2 - \alpha^2)\}^{-1}. \tag{1.73}$$

Provided therefore that $\Omega > \sqrt{2}\,\alpha$, we have

$$\Omega_m = (\Omega^2 - 2\alpha^2)^{1/2}$$

$$\simeq \Omega - \alpha^2/\Omega \qquad \text{if } \alpha \ll \Omega \tag{1.74}$$

and

$$A_m^2 = 1/4\alpha^2(\Omega^2 - \alpha^2)$$

$$\simeq 1/4\alpha^2\Omega^2 \qquad \text{if } \alpha \ll \Omega.$$

If, however, $\alpha > \Omega/\sqrt{2}$, A^2 has no maximum and decreases steadily to zero as ω increases. The ratio of the maximum to the zero frequency value is

$$A_m^2 \Omega^4 = \frac{\Omega^2}{4\alpha^2}\left(1 - \frac{\alpha^2}{\Omega^2}\right)^{-1} \tag{1.75}$$

which is very large for small damping. This describes *resonance*: large amplitudes are excited when a lightly damped oscillator is driven near its natural frequency.

The resonant efficiency of an oscillator is usually described in terms of the 'quality factor' Q. This is actually equal to $\Omega/2\alpha$, but is fundamentally defined as follows. The force $F \cos \omega t$ does work on the oscillator in order to drive it. The instantaneous power thus consumed is the product of force and resultant velocity $\dot{x}$

$$P(t) = F \cos \omega t\, \dot{x} = -F^2 A\omega \cos \omega t \sin(\omega t - \phi) \tag{1.76}$$

from (1.70). The *mean* power consumption $\bar{P}$ is obtained by averaging this over one cycle of oscillation

$$\bar{P} = \frac{\omega}{2\pi}\int_0^{2\pi/\omega} P(t)\,dt = \tfrac{1}{2}F^2 \omega A \sin \phi \tag{1.77}$$

$$= -\tfrac{1}{2}F^2\omega \operatorname{Im} W^{-1} \qquad \text{from (1.69)}$$

$$= F^2 \alpha \omega^2 [(\omega^2 - \Omega^2)^2 + 4\alpha^2\omega^2]^{-1} \qquad \text{from (1.65).} \tag{1.78}$$

This quantity is plotted against ω in figure 1.8. Unlike A^2 (figure 1.7) $\bar{P}$ is maximum exactly at $\omega = \Omega$, with value

$$\bar{P}_m = F^2/4\alpha. \tag{1.79}$$

Let ω_1 and ω_2 be those frequencies on either side of Ω for which $\bar{P} = \tfrac{1}{2}\bar{P}_m$. The quantity $\omega_2 - \omega_1$ is the power *bandwidth* of the oscillator. It is a measure of the range of frequencies between which power resonance is achieved by the driving force.

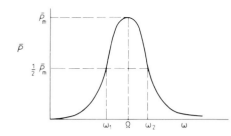

Figure 1.8 Mean power consumption of an oscillator.

The quality factor Q is *defined* as

$$Q = \Omega/(\omega_2 - \omega_1). \tag{1.80}$$

ω_1 and ω_2 are roots of the equation (see (1.78), (1.79))

$$\alpha\omega^2[(\omega^2 - \Omega^2)^2 + 4\alpha^2\omega^2]^{-1} = 1/8\alpha$$

from which it follows that

$$\omega_{1,2} = (\Omega^2 + \alpha^2)^{1/2} \mp \alpha$$

and so the power bandwidth is

$$\omega_2 - \omega_1 = 2\alpha \tag{1.81}$$

and

$$Q = \Omega/2\alpha \tag{1.82}$$

as stated. Notice that ω_1 and ω_2 are identical to the ω_-, ω_+ obtained in (1.72) from the phase variation.

All properties of free or forced oscillators can be described in terms of Q. For instance, the resonant/zero frequency amplitude ratio in equation (1.75) is $Q(1 - 1/4Q^2)^{-1/2} \simeq Q$ if Q is large enough. Again, the transient part of the solution becomes negligible after a time of order a few Q cycles of oscillation.

This concludes our discussion of individual oscillators. In Chapter 2, we shall consider systems of two or more *coupled* oscillators.

Normal Modes of Oscillation

2.1 Coupled Pendulums and Normal Modes of Oscillation

Isolated oscillators such as the simple pendulum are uncommon in nature. More often a system may consist of several parts, each capable of oscillating in its own right with its own frequency, but in which these parts can influence one another. New and unexpected phenomena arise in this situation, even when the system is linearised, and we shall investigate them by considering a simple, if artificial, example. This consists, as sketched in figure 2.1, of two pendulums moving in the same plane but coupled together by a spring which connects their bobs.

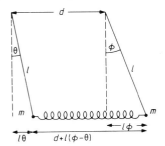

Figure 2.1 Coupled pendulums.

We shall assume that the two pendulums are of equal length and mass l and m, and that they are suspended a distance d apart. The spring has length d in its undeformed state. The simplest possible assumptions will be made about the system. From the outset the angles of inclination θ and ϕ will be assumed small enough to render linearisation an adequate approximation. The spring will be taken to have negligible mass, to always remain straight and to obey Hooke's law: the force T_s in it is proportional to its extension whether positive or negative. Thus if its length is x, the force is

$$T_s = K(x - d) \tag{2.1}$$

where K is the *spring constant*, and depends on its shape and the elastic properties of its material. The force is directed inwards (tension) if $x > d$ and outwards (compression) if $x < d$.

For given small values of the angles θ and ϕ, the length of the spring is approximately $d + l(\phi - \theta)$. So the force in it will be by equation (2.1), $Kl(\phi - \theta)$. If we resolve the forces of gravity and of the spring at right angles to the vertical, we find the linearised forms of the equations of motion for the two

$$ml\ddot{\theta} = -mg\theta + Kl(\phi - \theta) \qquad ml\ddot{\phi} = -mg\phi - Kl(\phi - \theta). \tag{2.2}$$

It is always worthwhile checking at this stage that the various signs make sense. Figure 2.1 shows that, when $\phi > \theta$, the spring is in tension, and so the spring acts so as to increase θ and decrease ϕ.

We now have a pair of second-order DEs (2.2) for θ and ϕ. These being two in number, we say that our system has two *degrees of freedom*. The equations are *coupled*. Because of the spring, each involves both θ and ϕ. Four arbitrary constants will occur in the general solution, and these will obviously correspond to the arbitrary specification of the angles θ and ϕ, and angular speeds $\dot{\theta}$ and $\dot{\phi}$ at the initial time. Before proceeding with the solution it is helpful (as it always is) to simplify the notation. The quantity $(g/l)^{1/2}$ is the frequency that each pendulum would have in the absence of the spring. This will no doubt have a role to play in the solution of this more general problem and we will, as in §1.2, call it Ω. Additionally, we introduce a quantity λ related to the spring constant K. Thus

$$\Omega = (g/l)^{1/2} \qquad \lambda = K/m. \tag{2.3}$$

Equations (2.2) become

$$\ddot{\theta} = -\Omega^2\theta + \lambda(\phi - \theta) \qquad \ddot{\phi} = -\Omega^2\phi - \lambda(\phi - \theta). \tag{2.4}$$

The obvious procedure would involve the elimination of either θ or ϕ to obtain an equation involving only one of these variables, but this is very cumbersome. Instead, an algebraic trick will be used. We notice that if we successively add and subtract the two equations we get

$$\ddot{\theta} + \ddot{\phi} = -\Omega^2(\theta + \phi) \qquad \ddot{\theta} - \ddot{\phi} = -(\Omega^2 + 2\lambda)(\theta - \phi). \tag{2.5}$$

The first of these involves only $\theta + \phi$, and the second $\theta - \phi$. Let us emphasise this by introducing two new variables, α and β

$$\alpha = \theta + \phi \qquad \beta = \theta - \phi \tag{2.6}$$

so that

$$\ddot{\alpha} = -\Omega^2\alpha \qquad \ddot{\beta} = -(\Omega^2 + 2\lambda)\beta. \tag{2.7}$$

These equations are uncoupled. The first involves α alone, and the second β. The general solution is immediate and simple harmonic

$$\alpha = \text{Re}(A\, e^{i\Omega t}) \qquad \beta = \text{Re}(B\, e^{i\Omega' t}) \tag{2.8}$$

where

$$\Omega' = (\Omega^2 + 2\lambda)^{1/2}. \tag{2.9}$$

Ω' is a frequency that involves both the natural frequency Ω of each pendulum and the spring constant λ. A and B are two arbitrary complex numbers that provide through their real and imaginary parts the four constants needed in the general solution.

Before solving an initial value problem we shall consider a pair of very special solutions that are of great importance. These have either A or B in equation (2.8) equal to zero.

(1) If B is zero, $\beta = 0$ and by equation (2.6)
$$\theta = \phi = \tfrac{1}{2}\,\mathrm{Re}\,(A\,e^{i\Omega t}). \tag{2.10}$$

(2) If A is zero, $\alpha = 0$ and
$$\theta = -\phi = \tfrac{1}{2}\,\mathrm{Re}\,(B\,e^{i\Omega' t}). \tag{2.11}$$

Two properties of these special solutions are obvious: (a) the variables θ and ϕ are very simply related and (b) each of them varies sinusoidally with the same frequency. This frequency is Ω in case (1) and Ω' in case (2).

Such motions occur in all linear coupled oscillatory systems. They are commonly called *normal modes of oscillation*. In a system with N degrees of freedom there would be (as we shall see in §2.2) N different normal modes each with its characteristic frequency. Here $N = 2$. Let us investigate the motions in the two modes. In mode (1), $\theta = \phi$ and the two pendulums oscillate together as sketched in figure 2.2(a). The spring retains its natural length throughout the motion. It has therefore no effect on the pendulums, and the result that each pendulum oscillates at the frequency Ω is quite natural. In mode (2), $\theta = -\phi$ and the behaviour is displayed in figure 2.2(b). Again we have synchronism, but the spring is now effective. Its extension and compression contribute with gravity to the restoring force, and inevitably therefore the frequency of oscillation will increase. We expect that $\Omega' > \Omega$, as is the case.

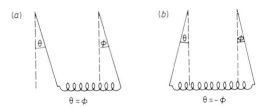

Figure 2.2 Normal modes of coupled pendulums.

The normal modes have been derived rather simply, merely by adding and subtracting equations (2.4). This is possible because of the assumed symmetry of the system: equal masses for the two pendulums. Otherwise it would not be clear how to proceed, and the linear combinations of θ and ϕ relevant to the modes would not be as simple as equation (2.6). The following procedure would, however, always work. Assume from the outset that in a normal mode each of θ

and ϕ oscillates with the same frequency, ω say

$$\theta = \text{Re}\,(R\,e^{i\omega t}) \qquad \phi = \text{Re}\,(S\,e^{i\omega t}). \qquad (2.12)$$

ω is to be determined together with some relation between the complex amplitudes R and S. Substituting equations (2.12) into (2.4) identities are obtained, provided that $-\omega^2 R = -\Omega^2 R + \lambda(S-R)$ and $-\omega^2 S = -\Omega^2 S - \lambda(S-R)$ or, collecting terms

$$(\omega^2 - \Omega^2 - \lambda)R + \lambda S = 0 \qquad \lambda R + (\omega^2 - \Omega^2 - \lambda)S = 0. \qquad (2.13)$$

Thus the ratio R/S can be written in two ways

$$R/S = -\lambda/(\omega^2 - \Omega^2 - \lambda) = -(\omega^2 - \Omega^2 - \lambda)/\lambda. \qquad (2.14)$$

Thus, an equation is obtained for the unknown frequency ω

$$(\omega^2 - \Omega^2 - \lambda)^2 = \lambda^2 \quad \text{so} \quad \omega^2 = \Omega^2 \quad \text{or} \quad \Omega^2 + 2\lambda. \qquad (2.15)$$

The two roots are precisely our mode frequencies, Ω and Ω'. Further, when $\omega^2 = \Omega^2$, equation (2.14) yields $R/S = 1$, and when $\omega^2 = \Omega^2 + 2\lambda$, it yields $R/S = -1$. The relation between the amplitudes in equations (2.10) and (2.11) is thus recovered. Had the masses of the bobs been different, we would have obtained for ω^2 a more intricate equation than (2.15) involving the ratio of these masses. Nevertheless, it would still be quadratic and we would obtain two modes.

Returning now to the general solution, we have immediately from equations (2.6) and (2.8)

$$\theta = \tfrac{1}{2}(\alpha + \beta) = \tfrac{1}{2}\,\text{Re}\,(A\,e^{i\Omega t} + B\,e^{i\Omega' t})$$
$$\phi = \tfrac{1}{2}(\alpha - \beta) = \tfrac{1}{2}\,\text{Re}\,(A\,e^{i\Omega t} - B\,e^{i\Omega' t}). \qquad (2.16)$$

This is just a special combination of the normal modes. The general solution is similarly expressible for any coupled linear system.

Consider a particular problem. When $t = 0$, the left-hand pendulum (θ) is released from rest with $\theta = \theta_0$, while the right-hand one (ϕ) is vertical and at rest. In the absence of the spring, the left would of course oscillate at Ω, while the right would remain at rest. What is the effect of the spring? The initial conditions are: $\theta = \theta_0$, $\dot{\theta} = 0$, $\phi = 0$, $\dot{\phi} = 0$ at $t = 0$. Substitution in equations (2.16) yields

$$\theta_0 = \tfrac{1}{2}\,\text{Re}\,(A+B) \qquad 0 = \tfrac{1}{2}\,\text{Re}\,[i(\Omega A + \Omega' B)]$$
$$0 = \tfrac{1}{2}\,\text{Re}\,(A-B) \qquad 0 = \tfrac{1}{2}\,\text{Re}\,[i(\Omega A - \Omega' B)] \qquad (2.17)$$

for the determination of the complex numbers A and B. It is easy to see that A and B have to be real and each equal to θ_0. Then our solution (2.16) becomes very simply

$$\theta = \tfrac{1}{2}\theta_0(\cos \Omega t + \cos \Omega' t) \qquad \phi = \tfrac{1}{2}\theta_0(\cos \Omega t - \cos \Omega' t). \qquad (2.18)$$

Notice the way in which the use of the complex exponential notation has rendered these manipulations very simple. In order to understand the content of equations (2.18) we use a pair of trigonometric identities

Normal Modes of Oscillation

$$\theta = \theta_0 \cos\left[\tfrac{1}{2}(\Omega' + \Omega)t\right] \cos\left[\tfrac{1}{2}(\Omega' - \Omega)t\right]$$
$$\phi = \theta_0 \sin\left[\tfrac{1}{2}(\Omega' + \Omega)t\right] \sin\left[\tfrac{1}{2}(\Omega' - \Omega)t\right].$$
(2.19)

These are most readily understood if we assume that the spring is *weak* in the sense that

$$\lambda \ll \Omega^2 \quad \text{or} \quad K \ll mg/l.$$
(2.20)

Then the frequency Ω' is approximately obtained from equation (2.9)

$$\Omega' \simeq \Omega + \lambda/\Omega$$

and so
$$\Omega' + \Omega \simeq 2\Omega \quad \text{and} \quad \Omega' - \Omega \simeq \lambda/\Omega.$$
(2.21)

Thus, approximately
$$\theta = \theta_0 \cos \Omega t \cos \gamma t \qquad \phi = \theta_0 \sin \Omega t \sin \gamma t$$
(2.22)

where
$$\gamma = \lambda/2\Omega \quad \text{and so} \quad \gamma \ll \Omega.$$
(2.23)

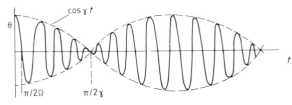

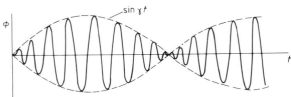

Figure 2.3 Beats.

θ and ϕ are sketched as functions of t in figure 2.3. Briefly, the rapid oscillations at the natural frequency Ω are *modulated* at the much lower frequency γ. These modulations are called *beats*. While $t \ll 1/\gamma$, θ oscillates at its natural frequency Ω, while ϕ remains small. But as t approaches $\pi/2\gamma$, the amplitude of the θ oscillations decreases to zero. Simultaneously, ϕ oscillates with increasing amplitude until at $t = \pi/2\gamma$, it has acquired all the original motion. The whole cycle is repeated periodically thereafter as is clear from the figure. No matter how weak is the spring (how small is γ/Ω) this phenomenon will occur if we wait long enough. The initial motion is transferred entirely to the pendulum originally at rest.

In the next section we consider a system with many degrees of freedom and corresponding modes.

2.2 The Normal Modes of a Stretched String Carrying N Masses

Having investigated coupled pendulums with two degrees of freedom we shall now study a particular system with N such degrees, where N is an arbitrary integer. Consider, as sketched in figure 2.4, a massless elastic string stretched between two fixed points A and B, and having attached to it N equal masses m at equal intervals a. The system performs transverse oscillations. The influence of gravity is neglected. Motions in the direction of the string are not considered. (Alternatively, the masses can be assumed to be constrained to move smoothly on straight wires perpendicular to AB.) The displacement of the rth mass will be called y_r, where r runs from 1 to N. This mass is acted on by the tensions in the strings attaching it to the $(r-1)$th and $(r+1)$th.

Figure 2.4 Masses displaced on a stretched string.

Let the undisturbed tension in the string be T_0. We assume that it obeys Hooke's law: its local extension is proportional to the tension. The tension in the string joining masses r and $r+1$ is thus $T_0 l/a$. The component of this normal to AB is $T_0(l/a)\sin\alpha = T_0(y_{r+1}-y_r)/a$. The corresponding expression for the string joining masses $r-1$ and r is $T_0(y_{r-1}-y_r)/a$. The equation of motion of the rth mass is then

$$m\ddot{y}_r = [T_0(y_{r+1}-y_r)/a] + [T_0(y_{r-1}-y_r)/a]$$

or

$$\ddot{y}_r = \Omega^2(y_{r+1} - 2y_r + y_{r-1}) \tag{2.24}$$

where

$$\Omega = (T_0/ma)^{1/2}. \tag{2.25}$$

Ω has the dimensions of a frequency. Equation (2.24) applies for $r=2$ to $r=N-1$. The boundary conditions for our system, the fact that the string is fixed at A and B, can be allowed for by introducing two extra dummy coordinates y_0 and y_{N+1}, each of which is zero. Then (2.24) will apply for $r=1$ and $r=N$ too.

Note that we have arrived without approximation at a *linear* set of equations. This circumstance derives from the use of Hooke's law, and is rather unusual. Almost always the analysis is akin to that for the simple pendulum: non-linear equations first arise and are then linearised for small amplitude motions.

There are two reasons for discussing this system. It can be solved *exactly* whatever N is. Following from this we can consider what happens when $N \to \infty$,

Normal Modes of Oscillation

while the individual masses and separations tend to zero. In this limit, a *continuous* distribution of mass is obtained; a system with an infinite number of degrees of freedom. As will be seen in §2.3, the subject of *waves* on a heavy elastic string can be approached in this fashion.

Equations (2.24) constitute a system of N coupled differential equations, to be compared with the two equations (2.4). A normal mode is, by definition, a motion in which each y_r performs SHM with the common frequency ω. Thus we substitute into (2.24)

$$y_r = \mathrm{Re}\,(Y_r\, e^{i\omega t}) \tag{2.26}$$

and see that

$$-\omega^2 Y_r = \Omega^2 (Y_{r+1} - 2Y_r + Y_{r-1})$$

or

$$Y_{r+1} - (2 - \omega^2/\Omega^2) Y_r + Y_{r-1} = 0 \tag{2.27}$$

for $1 \leqslant r \leqslant N$ with boundary conditions

$$Y_0 = Y_{N+1} = 0. \tag{2.28}$$

From this set of N simultaneous linear *algebraic* equations, we must extract a formula for the mode frequency ω and expressions for the ratios of the Y_r to one another. (Compare equations (2.13), (2.14) and (2.15) for the case $N = 2$.)

An equation connecting subscripted quantities such as (2.27) is called a *recurrence relation*. A linear one of this kind is known to be solvable using powers of a certain constant. Assume therefore that

$$Y_r \propto \lambda^r \tag{2.29}$$

where λ is to be determined. This will satisfy (2.27) *for all r* provided that

$$\lambda^2 - (2 - \omega^2/\Omega^2)\lambda + 1 = 0. \tag{2.30}$$

We shall assume that $\omega < 2\Omega$. In this case it is easy to see that the quadratic equation has complex conjugate roots of unit modulus. Indeed, if the real angle θ is introduced such that

$$\omega = 2\Omega \sin \tfrac{1}{2}\theta \qquad \text{for } (0 < \theta < \pi) \tag{2.31}$$

then (2.30) becomes

$$\lambda^2 - 2\cos\theta\,\lambda + 1 = 0 \tag{2.32}$$

whose roots are $\lambda_1 = e^{i\theta}$, $\lambda_2 = e^{-i\theta}$.

The *general* solution of (2.27) is

$$Y_r = A\lambda_1^r + B\lambda_2^r \tag{2.33}$$

where A and B are arbitrary constants. (There is a basic similarity to differential equation analysis here.) Inserting the values of λ_1, λ_2 we can write equivalently

$$Y_r = a\cos r\theta + b\sin r\theta. \tag{2.34}$$

The boundary condition $Y_0 = 0$ shows that $a = 0$. Then $Y_{N+1} = 0$ provided that

$$\sin(N+1)\theta = 0 \quad \text{or} \quad \theta = n\pi/(N+1) \qquad (2.35)$$

where n is any integer. The boundary conditions are thus satisfied only for a discrete set of values of θ. θ is restricted to lie between 0 and π, and so the values $1 \leq n \leq N$ will generate, via (2.31), just N different values of ω. *These are the normal mode frequencies.* Each of them is less than 2Ω, as was assumed in the analysis of (2.30). The boundary conditions cannot be satisfied if $\omega > 2\Omega$.

In summary therefore, the frequency of the nth mode is

$$\omega_n = 2\Omega \sin[n\pi/2(N+1)] \qquad (1 \leq n \leq N) \qquad (2.36)$$

and in this mode, the complex amplitude of the rth mass is

$$Y_r = b \sin(nr\pi/N+1) \qquad (1 \leq r \leq N) \qquad (2.37)$$

and

$$y_r = b \sin(nr\pi/N+1) \cos \omega_n t.$$

b is an arbitrary real measure of these amplitudes.

Consider for instance the case $N = 5$. The frequencies are $\omega_n = 2\Omega \sin(n\pi/12)$.

n	1	2	3	4	5
$\omega_n/2\Omega$	0.26	0.50	0.71	0.87	0.97

The five modes are sketched in figure 2.5. It is clear that for sufficiently large N, the various mode shapes approximate to sine *curves*. We anticipate that this behaviour will persist when $N \to \infty$.

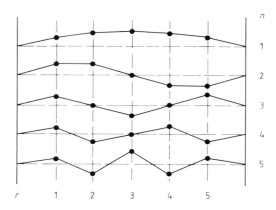

Figure 2.5 Normal modes for $N = 5$.

Before considering this limit, let us look briefly at another solution of the equation of motion which is *not* a normal mode.

Normal Modes of Oscillation

Consider (2.33). A particular solution (that does not satisfy the boundary conditions) is obtained with $A = 0$ and

$$Y_r = B \, e^{-ir\theta}. \tag{2.38}$$

The motion of the rth mass is then described (2.26) by

$$y_r = B \cos(\omega t - r\theta). \tag{2.39}$$

There is *no* restriction now on ω, which is related to θ through (2.31). The motions of the various masses are identical with a time delay between successive values of r. Thus y_2 repeats the motion of y_1 at a time θ/ω later, and so on. We have a *travelling* sinusoidal disturbance; our first example of a *wave motion*. The wave travels a distance a in a time θ/ω, and hence its speed V is

$$\begin{aligned} V = a/(\theta/\omega) &= a\Omega \sin(\tfrac{1}{2}\theta)/\tfrac{1}{2}\theta \\ &= (T_0 a/m)^{1/2} \sin(\tfrac{1}{2}\theta)/\tfrac{1}{2}\theta \end{aligned} \tag{2.40}$$

using (2.31) and (2.25). The importance of this expression will emerge in the next section.

2.3 The Limit $N \to \infty$ and the Continuous String

N oscillating masses have been considered in §2.2, and we now allow N to become large and a small in such a way that Na remains constant. The length L of the string is $(N+1)a$, which differs negligibly from Na as $a \to 0$. Concurrently, allow the individual mass m to become small in such a way that m/a, the mass on unit length of the string, remains constant. This will be denoted by σ. The distance ra of the rth mass from the left-hand end of the string becomes in the limit a *continuous* variable x. We shall therefore substitute

$$a = L/N \qquad m = \sigma L/N \qquad r = Nx/L \tag{2.41}$$

and let $N \to \infty$.

The amplitude y_r becomes a function $y(x, t)$ of x as well as t. We analyse the various findings of §2.2 in this limit.

Consider first the equation of motion (2.24). The frequency Ω (equation (2.25)) becomes

$$\Omega = N(T_0/\sigma)^{1/2}/L \tag{2.42}$$

which itself becomes infinite, but (2.24) is written

$$\partial^2 y/\partial t^2 = T_0 [y(x+a, t) - 2y(x, t) + y(x-a, t)]/ma. \tag{2.43}$$

Note that the term $\ddot{y}_r$ becomes a *partial* time derivative now, as y depends on x as well as t. As a tends to zero, we can expand the right-hand side by Taylor's theorem and obtain

$$\frac{\partial^2 y}{\partial t^2} = \frac{T_0}{ma}\left(y + a\frac{\partial y}{\partial x} + \tfrac{1}{2}a^2\frac{\partial^2 y}{\partial x^2} + \cdots\right.$$

$$-2y$$

$$\left. + y - a\frac{\partial y}{\partial x} + \tfrac{1}{2}a^2\frac{\partial^2 y}{\partial x^2} + \cdots\right)$$

$$= \frac{T_0 a}{m}\frac{\partial^2 y}{\partial x^2} + \cdots. \qquad (2.44)$$

t being fixed on the right-hand side, these derivatives are partial with respect to x. The arguments of each term are now x and t. Had not the terms in y and $\partial y/\partial x$ cancelled, there would be no proper limit because $1/ma$ and $1/m$ each tend to ∞. Terms involving higher derivatives than the second would have involved a^2/m, a^3/m etc, each of which tends to zero. Using (2.41), we see that our equation of motion becomes in the limit

$$\partial^2 y/\partial t^2 = c^2(\partial^2 y/\partial x^2) \qquad \text{where} \qquad c = (T_0/\sigma)^{1/2}. \qquad (2.45)$$

This is a *partial differential equation* (PDE) of the most fundamental importance, and will be studied in the remainder of the book. The quantity c has the dimension of *velocity*.

Consider next the normal modes. The boundary conditions are $y = 0$ at $x = 0$ and $x = L$ for all t. For given ω, (2.31) shows that θ tends to zero as $\Omega \to \infty$, and

$$\omega \simeq \Omega\theta. \qquad (2.46)$$

Equation (2.35) determining the frequencies is, in the limit,

$$\sin(N\omega/\Omega) = 0$$

or in view of (2.42) and (2.45)

$$\sin(\omega L/c) = 0. \qquad (2.47)$$

The modal frequencies are thus

$$\omega_n = cn\pi/L \qquad (2.48)$$

for *all* positive integers n: there is an *infinite* number of normal modes. Equation (2.48) also follows directly from (2.36). The displacement (2.37) becomes

$$y(x, t) = b \sin(n\pi x/L) \cos(n\pi ct/L) \qquad (2.49)$$

and it is easy to check that this satisfies (2.45) and the boundary conditions. This expression will be analysed in Chapter 3.

Finally the particular solution (2.39) (not a normal mode) becomes, with (2.46) and (2.41)

$$y = B \cos \omega(t - x/c). \qquad (2.50)$$

This represents a wave, continuous in x and sinusoidal in both x and t. The speed is simply c, and this also follows as the limit of (2.40).

In Chapter 3, these matters will be approached directly, by deriving (2.45) for a massive stretched string and analysing its solution from scratch.

Waves—General Properties and Waves on a Stretched String

3

3.1 General Remarks

It is rather difficult to devise a precise definition of a wave. In this section, some familiar properties of waves will be used to construct the most common type of equation used in their analysis. (In fact, this equation has been considered briefly in §2.3.) Because of their familiarity, it will be helpful to take as an example waves on the surface of water. (As we shall see in Chapter 5, however, such waves are frequently more complicated than discussed here.) A water wave is a moving disturbance in the elevation of the surface. Although the water itself moves very little, energy is transported by the wave from one place to another. A stone dropped into water imparts some energy to it at its point of impact. Some distance away, and at a later time, a body floating on the water is caused to oscillate.

Consider at time $t=0$ a disturbance y in the height of the surface, as shown in figure 3.1(a). It is specified as a function of horizontal position x by, say

$$y = F(x). \tag{3.1}$$

In this book, consideration is limited to one spatial dimension only. If the disturbance is to justify the name *wave* it will *move*. Generally, its shape will vary as it moves, but a simple possibility, which we limit ourselves to for the time being, is that its shape remains constant. Suppose then that the disturbance moves to the *right* with speed c. We can place the origin of x at the point of maximum displacement. Obviously, this point will have moved to the point $x = ct$ after time t. The shape being constant, the equation of the water height at time t will be simply

$$y = F(x - ct) \tag{3.2}$$

as shown in figure 3.1(b).

How could equation (3.2) have arisen as the solution of any problem? Physical principles will yield an *equation of motion*, presumably differential, as in Chapters 1 and 2. Here, however, the dependent variable y depends on two independent variables; time t and position x. We expect to obtain a PDE. Several examples will appear later. For the time being we proceed in the reverse direction and ask only the mathematical question: which PDE will yield (3.2) as its solution? It is clear that the particular function $F(x)$ specifying the initial shape of the disturbance should be arbitrary. The PDE should contain no reference to it. Hence we should attempt to construct a relation independent of F between the partial derivatives $\partial y/\partial t$ and $\partial y/\partial x$. If we let $\xi = x - ct$, so that $y = F(\xi)$, we have by the *chain rule* of partial differentiation

$$\frac{\partial y}{\partial x} = \frac{dF}{d\xi}\frac{\partial \xi}{\partial x} = \frac{dF}{d\xi} \qquad \frac{\partial y}{\partial t} = \frac{dF}{d\xi}\frac{\partial \xi}{\partial t} = -c\frac{dF}{d\xi}. \qquad (3.3)$$

So we can eliminate $dF/d\xi$ to obtain immediately

$$\partial y/\partial t + c(\partial y/\partial x) = 0. \qquad (3.4)$$

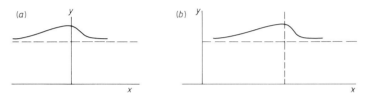

Figure 3.1 A moving disturbance of invariable form.

It is easy to show that equation (3.2) is the *general solution* of this. General solutions of PDEs involve *arbitrary functions* (in our case F) rather than the arbitrary constants involved in ordinary differential equations. The key property of equation (2.2) is not the function F but the dependence of its argument on x and t only through the combination $x - ct$.

Equation (3.4) is our first example of a *wave equation*. Its general solution represents a disturbance of arbitrary, but invariable, form travelling in the direction of positive x with speed c. This equation is of limited interest at any rate in mechanics. The reasons are twofold. Firstly we have singled out the *positive x* direction as that of travel, and there will usually be no reason for disturbances not to propagate equally well in the negative direction. Secondly equation (3.4) involves only the *first* time derivative of y. We shall always apply Newton's law in some form. Accelerations occur, and so *second* time derivatives are to be expected. Imagine equation (3.2) generalised to include two disturbances each of arbitrary shape propagating to the right and left with the same speed c. We would have

$$y = F(x - ct) + G(x + ct). \qquad (3.5)$$

This is illustrated in figure 3.2. The square wave F travels to the right, and the triangular one G to the left. Again we ask: which PDE will (3.5) satisfy for arbitrary shapes F and G? The number of arbitrary functions in the general solution of a PDE is equal to its order. We expect the latter therefore to be 2. Proceeding as before, we write $\xi = x - ct$, $\eta = x + ct$, and so

$$y = F(\xi) + G(\eta) \qquad \partial y/\partial x = F'(\xi) + G'(\eta) \qquad \partial y/\partial t = c(-F'(\xi) + G'(\eta)). \; (3.6)$$

Here F' and G' denote the respective ordinary derivatives. We can no longer derive a relation between $\partial y/\partial x$ and $\partial y/\partial t$ independent of both F and G. However, if we proceed to the second order

$$\partial^2 y/\partial x^2 = F''(\xi) + G''(\eta) \qquad \partial^2 y/\partial t^2 = c^2(F''(\xi) + G''(\eta)) \qquad (3.7)$$

and so we have the *second-order* PDE

$$\partial^2 y/\partial t^2 = c^2(\partial^2 y/\partial x^2). \qquad (3.8)$$

This has (3.5) as its general solution, valid for arbitrary functions F and G. Further, it involves the *second* time derivative $\partial^2 y/\partial t^2$, and so is capable, in principle, of encompassing accelerations. So both points mentioned above are met. We note that equations (2.45) and (3.8) are identical.

Figure 3.2 Waveforms moving simultaneously to right and left.

Equation (3.8), and its two- and three-dimensional generalisations, is of central importance in wave theory, and it arises in many (though by no means all) circumstances. Indeed it is commonly known as the *wave equation*. Ironically, as we shall see in Chapter 5, familiar water waves do not usually obey this equation!

In summary therefore, the wave equation (3.8) has as its general solution (3.5), which represents waves of arbitrary shape propagating simultaneously to the left and right with speed c. The two waves do not interact in any way. This is a consequence of the *linearity* of (3.8): the dependent variable y appears through its derivatives only to the first power.

In the remainder of this chapter we shall consider waves that can travel on a stretched string. This topic was considered briefly in §2.3, but will here be approached from first principles. Stringed musical instruments illustrate this topic, although the manner in which these vibrations generate associated sound waves is far beyond the scope of this book. Sound waves themselves are discussed in Chapter 4.

3.2 The Stretched String

A string stretched between two points vibrates if pulled aside and released. The vibrations can be heard if the string is tight enough. Various forms of vibration can be set up by plucking the string in different ways: applying different initial conditions.

The inertia of the string is opposed by its tension. Gravity would be of importance for exceedingly long 'strings' (e.g. suspension bridge cables) but is not considered here. The equation of motion is obtained by equating force to mass × acceleration, but the derivation is inevitably more complicated than it is for the pendulum.

When the string is undistorted, let it have mass per unit length σ, and tension T_0. We shall assume that it obeys Hooke's law, namely that the tension in a given section is proportional to the extension of that section. Figure 3.3 illustrates a portion of the string when pulled aside from its original line AB. A section PQ originally lies between x and $x + \Delta x$, relative to an origin on the line AB. When displaced it moves to P'Q', and is then of length Δl. P' lies at a distance y from AB. y will be our dependent variable, a function of x and of time t. Similarly Q' is at a distance $y + \Delta y$. The tension in P'Q' is by Hooke's law, $T_0 \Delta l/\Delta x$. If α is the angle made by the tangent to the string at P', then the component of the tension toward AB at P' is $T_0(\Delta l/\Delta x) \sin \alpha = T_0 \Delta y/\Delta x$. P'Q' is slightly curved, and the forces of tension at P' and Q' do not quite balance. The net outward force on P'Q' is

$$T_0[(\Delta y/\Delta x)_{x+\Delta x} - (\Delta y/\Delta x)_x] = T_0 \Delta x(\partial^2 y/\partial x^2) \qquad (3.9)$$

to the first order in Δx. The derivative is partial with respect to x because no time variations are yet considered.

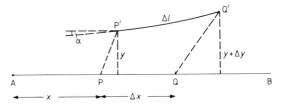

Figure 3.3 A section of the stretched string.

The mass of the section PQ or P'Q' is $\sigma \Delta x$. Its outward acceleration is $\partial^2 y/\partial t^2$. By Newton's law and (3.9) we have therefore

$$\sigma \Delta x \, \partial^2 y/\partial t^2 = T_0 \Delta x(\partial^2 y/\partial x^2) \qquad (3.10)$$

or

$$\partial^2 y/\partial t^2 = c^2(\partial^2 y/\partial x^2) \qquad (3.11)$$

where

$$c = (T_0/\sigma)^{1/2}. \qquad (3.12)$$

This is precisely our wave equation (3.8) and is identical to (2.45). The wave speed c is expressed in terms of the tension and density of the string when undistorted. For instance, if $\sigma = 0.05$ kg m^{-1} and $T_0 = 50$ kg wt $= 500$ N, then $c = 100$ m s^{-1}. It is intuitively reasonable that c should increase with T_0 and decrease with σ. It has *not* been assumed that the displacement y or the angle α are small. Nevertheless, a *linear* equation has been derived. This is rare. Almost always, as we shall see, wave equations derived initially are *non-linear*, and appropriate linearisation is carried out in the first stages of an investigation. Most treatments of our problem assume from the outset that the angle α is very small, and that the tension T_0 is so large that it can be taken to be constant. The assumed validity of Hooke's law here renders such approximations unnecessary, as was also the case in §2.2. (Of course, if the string were stretched too far this law would fail, and we would then have a non-linear problem.)

3.3 Waves and Oscillations on a Stretched String

We already know from §3.1 the general solution of our problem: equation (3.5), representing arbitrary waveforms travelling to left and right with speed c. What then remains to be considered? In practice the string is of bounded length. At a fixed end the displacement y is zero for all time. This provides an all important *boundary condition* for problems. The process of adjusting the solution of a PDE to satisfy given boundary conditions is akin to the calculation of the constants occurring in the solution of a DE. It is, however, usually rather an intricate procedure. We shall consider two problems of this nature.

3.3.1 Reflection of a Wave from a Fixed Point

A pulse (a wave of limited spatial extent) can be induced in a string by fixing one end and flicking the other end with the hand. It travels away from the hand, is *reflected* at the fixed end, and returns to the hand. The manner in which this phenomenon is analysed follows.

In figure 3.4(a) the pulse travels toward the left. The string is fixed at $x=0$. Until the pulse arrives at $x=0$ it is given by

$$y = f(x+ct) \tag{3.13}$$

where the function $f(x)$ describes its shape. To ensure, however, that $y=0$ for all t when $x=0$, (3.13) must be augmented. This is accomplished by the addition of a wave, of an as yet unknown form, travelling to the right. It is clear from the general solution (3.5) that this is the only possibility. Thus

$$y = f(x+ct) + g(x-ct). \tag{3.14}$$

The boundary condition $y=0$ at $x=0$ for all t gives $f(ct)+g(-ct)=0$ which determines $g(\xi)$ in terms of $f(\xi)$ for all values of the argument ξ

$$g(\xi) = -f(-\xi). \tag{3.15}$$

The solution is therefore from (3.14)

$$y = f(x+ct) - f(ct-x). \tag{3.16}$$

The second term is a pulse *reflected* from the fixed end. It is similar to the incident pulse, but is of opposite sign and is reversed in space. Three stages in the reflection are sketched in figure 3.4 for an unsymmetrical pulse. A very helpful construction is also shown. Imagine a 'ghost' string extending to the left from $x=0$. The wave $-f(ct-x)$ is incident from the left. At each instant t, this is added appropriately to the $f(x+ct)$, and the portion in $x<0$ discarded.

3.3.2 Sinusoidal Waves, Standing Waves, and Normal Modes

So far, sines and cosines have not entered our considerations. They have a vital role to play in wave theory, and appear in the following way. The wave equation

$$\partial^2 y/\partial t^2 = c^2(\partial^2 y/\partial x^2) \tag{3.17}$$

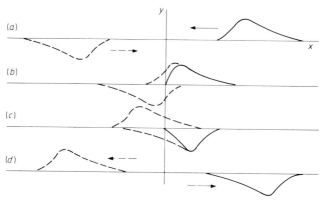

Figure 3.4 Reflection of a pulse.

has (3.5) as general solution. The latter is not always convenient. Let us investigate instead solutions which factorise into a function of x multiplied by one of t. Assume then that

$$y = f(x)g(t) \tag{3.18}$$

and substitute this into equation (3.17). After dividing by fg, we get

$$\frac{1}{g}\frac{d^2 g}{dt^2} = c^2 \frac{1}{f}\frac{d^2 f}{dx^2}. \tag{3.19}$$

The two sides of this relation are functions respectively of t and x alone, and it can be satisfied for all values of t and x only if each side is *constant*. Writing this constant as $-\omega^2$

$$d^2 g/dt^2 = -\omega^2 g \qquad d^2 f/dx^2 = -\omega^2 f/c^2. \tag{3.20}$$

Each of these possesses sinusoidal solutions. Had we chosen a positive constant, exponential solutions would have arisen. These are sometimes necessary, but are not considered here. In particular, we can take

$$g = A \cos \omega t \qquad f = B \cos \omega x/c \tag{3.21}$$

where A and B are constants. Hence

$$y = 2C \cos \omega t \cos (\omega x/c) \tag{3.22}$$

where $C = \frac{1}{2}AB$. This is consistent with the general solution, as can be seen by writing it equivalently as

$$y = C\{\cos [\omega(x - ct)/c] + \cos [\omega(x + ct)/c]\} \tag{3.23}$$

which is of the form (3.5) in which the general functions F and G are cosines. The

Waves—General Properties and Waves on a Stretched String

two special travelling waves in equation (3.23) are naturally called *sinusoidal waves*. We consider briefly that wave travelling to the right (which is itself a particular solution of the wave equation) in order to define various important parameters associated with it. It can be written

$$y = C \cos(kx - \omega t). \tag{3.24}$$

$k = \omega/c$ is the *wave number*.

The spatial period over which y repeats itself is $\lambda = 2\pi/k = 2\pi c/\omega$. This is the *wavelength* and ω is the *angular frequency*.

The number of oscillations per second at a fixed point is $f = \omega/2\pi$, the *frequency* (measured in Hz). The *temporal period* is $T = 1/f = 2\pi/\omega$. The *amplitude* is C. Note that the wave speed c is given by ω/k or by $f\lambda$.

The solution (3.23) is a superposition of two sinusoidal waves identical but for their directions of propagation. The spatial form is sketched in figure 3.5. For all values of t, y vanishes when $x = \pm\pi c/2\omega = \pm\frac{1}{4}\lambda$, and periodically thereafter at the interval $\frac{1}{2}\lambda$. The amplitude of the spatial pattern itself varies sinusoidally as $2C \cos \omega t$. Such a disturbance does not therefore propagate and is known as a *standing wave*.

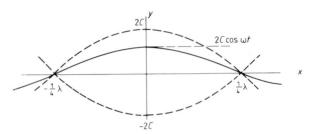

Figure 3.5 A standing wave.

Since standing waves vanish for all t at particular values of x they can be used to construct solutions for a string fixed at both ends. They are a good deal more convenient to use in this case than the general solution. Consider a string of length L fixed ($y = 0$) at $x = 0$ and $x = L$. Our standing wave (3.22) does not vanish at $x = 0$, but it will if replaced by

$$y = A \cos \omega t \sin kx \qquad (k = \omega/c). \tag{3.25}$$

If y is also to vanish at $x = L$ for all t, we need $\sin kL = 0$ and so the wave number k, so far unrestricted, takes *one of a discrete set of values*, $k = n\pi/L$, where n is any integer.

Correspondingly

$$\lambda = 2L/n \qquad \omega = n\pi c/L \qquad f = nc/2L \tag{3.26}$$

and such relations have also been defined in §2.3 for N masses attached to a weightless string, as $N \to \infty$.

The reason for this is clear. Figure 3.6 illustrates the displacement for $n = 1, 2$ and 3. Since y vanishes at $x = 0$ and $x = L$, an integral number of half wavelengths must be fitted between these points. The solutions (3.25) for these values of k and ω are called the *normal modes of vibration* of the string. In such a mode, each part of the string oscillates in phase at the particular frequency (3.26). This is analogous to the behaviour of coupled pendula in Chapter 2.

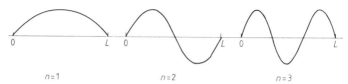

Figure 3.6 The first three normal modes for a bounded string.

The smallest frequency corresponds to $n = 1$, and is

$$f_1 = c/2L. \tag{3.27}$$

This is the *fundamental* frequency. For example, if $L = 1$ m, and $c = 100$ m s^{-1} then $f_1 = 50$ Hz. The frequencies of higher-order modes are in this case multiples of f_1 and are called *harmonics*.

How could a single normal mode be excited? As in Chapter 1, special initial conditions are necessary. For the nth mode

$$y = A \cos (n\pi ct/L) \sin (n\pi x/L) \tag{3.28}$$

and so at $t = 0$

$$y = A \sin (n\pi x/L) \quad \text{and} \quad \partial y/\partial t = 0. \tag{3.29}$$

To excite the mode (3.28) we should therefore have to displace the string by the amount (3.29) and release it from rest. By including a term in $\sin (n\pi ct/L)$ we could generalise to non-zero initial velocity, but for simplicity this will be omitted.

It would be rather difficult to do this. Suppose instead that the string is simply plucked and released at $t = 0$. The initial displacement can be quite general

$$y(x, 0) = F(x) \qquad \partial y(x, 0)/\partial t = 0. \tag{3.30}$$

$F(x)$ can be an arbitrary function satisfying $F(0) = F(L) = 0$.

In Chapter 2, the general motion of coupled pendulums was expressed as a sum of normal modes. A general sum of modes here could, by (3.28), take the form

$$y = \sum_{m=1}^{\infty} A_n \cos (n\pi ct/L) \sin (n\pi x/L). \tag{3.31}$$

For any values of the amplitudes A_n, this is obviously a solution of the wave equation that satisfies the boundary conditions. Can we choose the A_n in order

Waves—General Properties and Waves on a Stretched String

that the initial conditions (3.30) are satisfied? (3.31) already has $\partial y/\partial t = 0$ at $t = 0$, and so we would need the single condition

$$F(x) = \sum_{n=1}^{\infty} A_n \sin(n\pi x/L). \tag{3.32}$$

It can be shown, though it is beyond the scope of this book, that any function $F(x)$ vanishing at $x = 0$ and $x = L$ can be expressed in this form, and that the amplitudes A_n are readily determined from $F(x)$. This is the subject of *Fourier analysis*, and is of fundamental importance in wave theory. It follows that any motion of the bounded string is expressible as a sum of normal modes.

All bounded systems satisfying the wave equation possess normal modes, and the foregoing remarks apply. There is, however, one special feature of our example which must be noted. As we have seen, the frequencies of the modes are integral multiples of the fundamental. A violin or guitar string owes much of its musical tonality to this fact. In general, however, the higher frequencies are not as simply related to the fundamental. An example from the theory of sound will be described in the next chapter. The two-dimensional equivalent of the string, namely the *drum* can be shown to be another example. Its modal frequencies are not simply related, and it achieves its characteristic sound in large part from this fact.

3.4 Energetics

Finally, we consider the balance of energies for the string. For simplicity, discussion is limited to waves of small amplitude. The section PQ of the string in figure 3.3 has mass $\sigma \Delta x$, and speed $\partial y/\partial t$. Its kinetic energy is then $\frac{1}{2}\sigma \Delta x(\partial y/\partial t)^2$. The total for the section between $x = a$ and b is

$$K = \frac{1}{2}\sigma \int_a^b (\partial y/\partial t)^2 \, dx. \tag{3.33}$$

The work done in stretching the string from Δx to Δl is, if $\Delta l/\Delta x$ is close to unity, $T_0(\Delta l - \Delta x)$. For small displacements

$$\Delta l = (\Delta x^2 + \Delta y^2)^{1/2} = \Delta x[1 + (\partial y/\partial x)^2]^{1/2}$$
$$\simeq \Delta x[1 + \tfrac{1}{2}(\partial y/\partial x)^2]. \tag{3.34}$$

The potential energy of the section is therefore

$$P = \tfrac{1}{2}T_0 \int_a^b (\partial y/\partial x)^2 \, dx. \tag{3.35}$$

For a wave $y = F(x - ct)$ progressing in one direction, we have, on using the relation $c^2 = T_0/\sigma$

$$K = P = \tfrac{1}{2}T_0 \int_a^b [F'(x - ct)]^2 \, dx. \tag{3.36}$$

So kinetic and potential energies are equal. If a and b are fixed points, K and P will vary with time. However, if we allow a and b to move with the wave speed c, so that $a = A + ct$ and $b = B + ct$, then K and P are *constant*

$$K = P = \tfrac{1}{2} T_0 \int_A^B (F'(x))^2 \, dx. \tag{3.37}$$

In this sense we can speak then of the propagation of the constant energy with the speed c of the wave. For sinusoidal waves of form

$$y = C \cos [\omega(x - ct)/c] \tag{3.38}$$

the energies *per wavelength* are obtained with $A = 0$, $B = \lambda = 2\pi c/\omega$

$$K = P = \tfrac{1}{2}\pi C^2 T_0 k = \pi^2 C^2 T_0/\lambda. \tag{3.39}$$

If, for instance, $T_0 = 500$ N, $C = 0.01$ m, $\lambda = 0.1$ m, $K = P = 5$ J.

The balance between K and P is more complicated if waves travel in both directions. However, for the normal modes on a string of length L (equation (3.28)) it is readily shown that

$$K = \tfrac{1}{4} T_0 [(n\pi A)^2/L] \cos^2 (n\pi ct/L)$$

$$P = \tfrac{1}{4} T_0 [(n\pi A)^2/L] \sin^2 (n\pi ct/L). \tag{3.40}$$

The sum of K and P is constant, and the energy oscillates between K and P twice per cycle.

Sound Waves 4

4.1 The Mechanism of Sound

Despite the familiarity of sound, its manner of propagation is not obvious. However, it is quite easy to understand roughly why a gaseous medium should be capable of supporting waves. Quantitative application of these principles will follow in §4.2.

Air and gases in general are composed of *molecules*. These move very rapidly and undergo exceedingly frequent collisions with each other. Each of these results in an exchange of momentum between two molecules. The effect of many such collisions is the exertion of a force, the *pressure* between adjacent parts of the gas. It exists throughout a volume of gas and acts also on the containing walls. In a uniform region of gas, the *density* (mass of molecules per unit volume) and *temperature* (a measure of the mean speed of the molecules) are constant, and the pressure is in balance in all directions. The gas is in equilibrium. Non-uniformity will, however, result in a net force on the relevant part of the gas. The inertia of the particles opposes the tendency of this pressure gradient to restore equilibrium. The two essential ingredients for oscillations and waves are therefore present.

Equations governing these *sound waves* will be derived in the next section, in one spatial dimension. These are non-linear. There is scope in this book only to analyse the linearised versions of these equations.

4.2 The Wave Equation for Sound

Consider a straight pipe of cross-sectional area A containing gas (figure 4.1). We discuss only motions of the gas along the pipe in the x direction. The behaviour of the reference volume lying between the two fixed planes at x and $x + \Delta x$ is discussed. Δx is assumed small and will eventually tend to zero. This is *not* a fixed mass of gas, since motions occur across these two planes. The gas is characterised by its density ρ, velocity u and pressure p, each of which is a function of x and time t. Variations with the other transverse dimensions are ignored. We shall need three relations between these quantities.

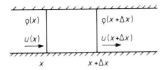

Figure 4.1 A reference volume of gas.

4.2.1 Conservation of Mass

The mass of gas in the reference volume is $\rho A \Delta x$. In time Δt this increases by $(\partial \rho/\partial t) A \Delta x \Delta t$ to first order in Δt. This must be balanced by the flows of gas across the ends. The *inflow* at x during this time is $\rho u A \Delta t$ evaluated at x. The *outflow* at $x + \Delta x$ is $\rho u A \Delta t$ evaluated at $x + \Delta x$. Hence

$$(\partial \rho/\partial t) A \Delta x \Delta t = [(\rho u)_x - (\rho u)_{x+\Delta x}] A \Delta t.$$

Divide by $A \Delta x \Delta t$ and take the limit as $\Delta x \to 0$. Thus

$$\partial \rho/\partial t = -\partial(\rho u)/\partial x. \tag{4.1}$$

This is the expression of *conservation of mass*, and is also called simply the *continuity* equation.

4.2.2 Conservation of Momentum

The momentum in the positive x direction of the gas in the reference volume is $\rho u A \Delta x$. In time Δt, this increases by $(\partial(\rho u)/\partial t) A \Delta x \Delta t$. This is balanced partly by flow of momentum across the ends, and partly by the difference in pressure of the fluid adjacent to the two ends. The inflow of momentum in time Δt across x is $(\rho u) u A \Delta t$ evaluated at x. A corresponding expression gives the outflow at $x + \Delta x$. Pressure being a force per unit area normal to the ends, the force on x in the positive x direction is pA evaluated at x. The force at $x + \Delta x$ in the negative direction is pA evaluated at $x + \Delta x$. The corresponding impulses are each Δt times these forces. On equating net change in momentum to net impulse, we get

$$(\partial(\rho u)/\partial t) A \Delta x \Delta t = [(\rho u^2)_x - (\rho u^2)_{x+\Delta x}] A \Delta t + (p_x - p_{x+\Delta x}) A \Delta t.$$

In the limit as $\Delta x \to 0$ this becomes

$$\partial(\rho u)/\partial t = -\partial(\rho u^2)/\partial x - \partial p/\partial x \tag{4.2}$$

which is the expression of *conservation of momentum*, and is called the Euler equation of motion after its originator.

4.2.3 Conservation of Energy: the Adiabatic Relation

A third relation is needed between our variables ρ, u and p. The principle of conservation of energy provides this via thermodynamics but its derivation is beyond the scope of this book. In his original investigations of sound, Newton assumed that the fluctuations would occur isothermally (at constant temperature). This would imply via Boyle's and Charles' laws that the pressure is

Sound Waves

directly proportional to the density. It was discovered experimentally that this yielded too small a value for the speed of sound and it is now known that, for sound waves of nearly all frequencies of interest, fluctuations occur so rapidly that no exchange of heat is possible between adjacent masses of gas. The temperature does fluctuate. Texts on thermodynamics show that the consequence of this is a relation berween pressure and density, the *adiabatic* law

$$p = k\rho^{\gamma}. \qquad (4.3)$$

We shall take k to be a constant for a given gas, γ is the ratio of specific heats at constant pressure and volume respectively. Again, it will be assumed constant for a given gas. Its value for air at STP is very nearly 7/5. We shall adopt the law (4.3) without further comment.

Our equations (4.1), (4.2) and (4.3) are now grouped together for convenience. If we expand the partial derivatives in (4.2) and use (4.1), some simplification follows. Additionally, let p_0 and ρ_0 be the values of p and ρ when the gas is undisturbed. k in (4.3) can then be expressed as $p_0\rho^{\gamma}$. Hence

$$\partial\rho/\partial t + \partial(\rho u)/\partial x = 0$$
$$\partial u/\partial t + u(\partial u/\partial x) = -\rho^{-1}(\partial p/\partial x) \qquad p/p_0 = (\rho/\rho_0)^{\gamma}. \qquad (4.4)$$

It is not obvious that a wave equation is hidden here! We shall henceforth consider only *linearised* sound waves.

4.3 Linearised Sound Waves

We shall linearise equations (4.4) directly. Let

$$\rho = \rho_0 + \rho_1 \qquad u = u_1 \qquad p = p_0 + p_1 \qquad (4.5)$$

where ρ_1 and p_1 are assumed much smaller than ρ_0 and p_0, and u_1 is small. Substitute into (4.4) and neglect products and squares of small quantities

$$\partial\rho_1/\partial t + \rho_0(\partial u_1/\partial x) = 0$$
$$\partial u_1/\partial t + \rho_0^{-1}(\partial p_1/\partial x) = 0 \qquad p_1/p_0 = \gamma\rho_1/\rho_0. \qquad (4.6)$$

Eliminating ρ_1 and p_1 we get our familiar linear wave equation

$$\partial^2 u_1/\partial t^2 = c^2(\partial^2 u_1/\partial x^2) \qquad (4.7)$$

where

$$c = (\gamma p_0/\rho_0)^{1/2} \qquad (4.8)$$

and ρ_1 and p_1 also satisfy this equation.

Some comments are necessary before proceeding with solutions. For air at STP we have $\rho_0 = 1.29$ kg m^{-3}, $p_0 = 1.007 \times 10^5$ Pa, yielding with $\gamma = 1.4$, $c = 331$ m s^{-1}, in accord with measurements. Newton's isothermal value is $(p_0/\rho_0)^{1/2}$ which is about 20% too low.

When is linearisation permissible? Sound pressure is usually measured in *decibels* (dB). This measure is given by $20 \log_{10}(p_1/p_{ref})$, where $p_{ref} = 2 \times 10^{-5}$ Pa corresponds roughly to the threshold of human hearing at a frequency of 1 kHz. A sound of 140 dB which is on the threshold of pain, has therefore $p_1 = 10^7 p_{ref} = 200$ Pa. Hence, $p_1/p_0 = 2 \times 10^{-3}$, a very small number! For nearly all practical purposes, linearised sound theory is adequate.

All linearised sound waves propagate with the same speed c. In Chapter 5 we shall meet an example of waves whose speed depends on their frequency. Such waves are called *dispersive* for the following reason. Any signal can be expressed (via Fourier analysis, as mentioned in Chapter 3) as a superposition of sinusoidal waves, each with a unique frequency. In a dispersive situation each of these would travel at a different speed. Their superposition at a distant point would then be distorted. The signal becomes *dispersed*, in the usual sense of the word. It is obviously vital for spoken communication that sound waves are not dispersive!

Most sound phenomena of interest are three-dimensional and beyond the scope of this book. Our one-dimensional wave equation is capable of describing, to a limited extent, the behaviour of an organ pipe. It will be used to illustrate some further elementary properties of waves.

4.4 Normal Modes of a Pipe

Consider a pipe of length L which is closed at one end, $x=0$, and open at the other, $x=L$. This is a very rough model of an organ pipe. The air remains in contact with the closed end, so one boundary condition is that $u_1 = 0$ at $x=0$. At the open end, the exact condition is very complicated. It can be approximated by the requirement that the pressure there is equal to the ambient pressure p_0. This clearly cannot be quite correct, since no sound would emerge from the pipe! But it is an adequate approximation if the pipe radius is much less than its length. It follows that $p_1 = 0$ at $x=L$ and so from equations (4.6), that ρ_1, $\partial \rho_1/\partial t$ and $\partial u_1/\partial x = 0$ for all t.

Separable solutions of (4.7) that satisfy $u_1 = 0$ at $x=0$ take the form, following Chapter 3

$$u_1 = C \cos \omega t \sin (\omega x/c). \tag{4.9}$$

This will satisfy also $\partial u_1/\partial x = 0$ at $x=L$, if

$$\cos (\omega L/c) = 0 \tag{4.10}$$

or

$$\omega = (n + \tfrac{1}{2})\pi c/L \tag{4.11}$$

where n is any integer. With these values of ω (4.9) gives the *normal modes* of the pipe with boundary conditions. The fundamental frequency is $\omega_0 = \pi c/2L$, $f_0 = c/4L$ which is 83 Hz for a one metre pipe. Higher modes have frequencies

Sound Waves

$(2n+1)\omega_0$, i.e. *odd* multiples of the fundamental. The first three modes are sketched in figure 4.2, and it is clear that the boundary conditions imply that an odd number of quarter wavelengths must fit between $x=0$ and L. Since $\lambda = 2\pi c/\omega$, this yields (4.11). The relation between the fundamental and higher frequencies is responsible to a limited extent for the tonal quality of organ pipes and other wind instruments.

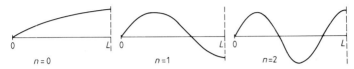

Figure 4.2 The first three normal modes for a pipe.

4.5 Intensity of Sinusoidal Sound Waves

The instantaneous *intensity* of a sound wave is given by the rate of working of the force of pressure. It is measured in W/m² and is given by $pu = (p_0 + p_1)u_1$. Consider a travelling sound wave whose pressure is given by

$$p_1 = P \cos \omega(t - x/c). \tag{4.12}$$

By (4.6) the velocity is given by

$$u_1 = (P/\rho_0 c) \cos \omega(t - x/c). \tag{4.13}$$

The intensity *averaged in time* over a cycle of the wave is given by

$$I = \omega/2\pi \int_0^{2\pi/\omega} p_1 u_1 \, dt = P^2/2\rho_0 c. \tag{4.14}$$

Note that the term $p_0 u_1$, involving the ambient pressure, has zero average. The decibel measure of this is $10 \log_{10} (I/I_{\text{ref}})$ where $I_{\text{ref}} = p_{\text{ref}}^2/\rho_0 c \simeq 10^{-12}$ W m⁻². A wave of 140 dB intensity has $I = 100$ W m⁻².

4.6 Reflection of Sound Waves

Consider a pipe of indefinite length, containing, at $x=0$, a plane element of mass M, free to move along the pipe without friction. A sinusoidal wave is incident on the element from $x<0$. Its mass influences the balance of forces at $x=0$, with the result that secondary waves are transmitted into $x>0$ and reflected back into $x<0$. These have the same frequency ω as the incident wave, but different amplitudes and phases. Referring to figure 4.3, let the complex amplitudes of the incident, reflected and transmitted pressure waves be unity, R and T respectively.

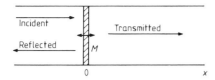

Figure 4.3 Reflection of waves from a movable mass M.

The wave pressures in $x < 0$ and $x > 0$ are

$$p_1 = \operatorname{Re} (e^{i\omega(t - x/c)} + R\, e^{i\omega(t + x/c)}) \qquad x < 0$$
$$p_1 = \operatorname{Re} (T\, e^{i\omega(t - x/c)}) \qquad x > 0. \tag{4.15}$$

The corresponding velocities follow from (4.6) and are

$$u_1 = \operatorname{Re} (1/\rho_0 c)(e^{i\omega(t - x/c)} - R\, e^{i\omega(t + x/c)}) \qquad x < 0$$
$$u_1 = \operatorname{Re} (1/\rho_0 c)\, e^{i\omega(t - x/c)} \qquad x > 0. \tag{4.16}$$

The boundary conditions at $x = 0$ are two-fold. The velocity must be the same for all times on the two sides of the mass. The difference in pressure across it must yield mass × acceleration. The acceleration is the same as that of the gas on each side: $\partial u_1/\partial t$. The first condition yields from (4.16)

$$1 - R = T. \tag{4.17}$$

The second can be written

$$A(p(x > 0) - p(x < 0))_{x=0} = M(\partial u_1/\partial t)_{x=0} \tag{4.18}$$

which reduces using (4.15) and (4.16) to

$$A(T - 1 - R) = Mi\omega T/\rho_0 c. \tag{4.19}$$

Equations (4.17) and (4.19) determine the amplitudes T and R. Introduce a dimensionless quantity μ proportional to the mass

$$\mu = M\omega/A\rho_0 c = Mk/A\rho_0 \tag{4.20}$$

where $k = \omega/c$ is the wave number. Then we have

$$R + T = 1 \qquad R - T(1 - i\mu) = -1 \tag{4.21}$$

with solution

$$R = -\tfrac{1}{2}i\mu/(1 - \tfrac{1}{2}i\mu) \qquad T = 1/(1 - \tfrac{1}{2}i\mu). \tag{4.22}$$

R and T are generally *complex*, and so both the amplitudes and phases of the reflected and transmitted waves are affected. For zero mass, $\mu = 0$, $R = 0$, $T = 1$, as of course expected. For large mass, $\mu \to \infty$, $R \to 1$ and $T \to 0$. A rigid wall reflects perfectly. As an example, consider a pipe of area $A = 10\,\text{cm}^2$ and a frequency of 1 kHz. k is then about $20\,\text{m}^{-1}$ and $\rho_0 \sim 1\,\text{kg m}^{-3}$. Thus $\mu \sim 2 \times 10^4\, M$. So for masses much larger than 0.05 g, μ is large. For this mass, μ is small for $f \ll 1$ kHz, and large if $f \gg 1$ kHz.

Sound Waves

The analysis is simplified by the introduction of an acute angle α in place of μ via $\mu = 2 \tan \alpha$. Thus

$$R = -i \sin \alpha \, e^{i\alpha} \qquad T = \cos \alpha \, e^{i\alpha} \tag{4.23}$$

and the waveforms (4.15) become

incident $\quad \cos[\omega(t - x/c)]$

reflected $\quad \sin \alpha \sin[\omega(t + x/c) + \alpha]$ \hfill (4.24)

transmitted $\quad \cos \alpha \cos[\omega(t - x/c) + \alpha]$.

So α is the phaseshift of the transmitted wave relative to the incident, and $\cos \alpha$ is its amplitude.

Notice finally that

$$|R|^2 + |T|^2 = 1. \tag{4.25}$$

Analysis similar to that in §4.5 shows that this expresses conservation of energy. The energies per wavelength in the respective waves are proportional to the squared moduli of their complex amplitudes. A fraction $\sin^2 \alpha$ of the incident energy is reflected, and a fraction $\cos^2 \alpha$ transmitted.

4.7 The Normal Modes of a Pipe with a Movable End

This final example from the theory of sound combines features of both previous ones, and is intended to illustrate techniques. A pipe of length L (figure 4.4) is closed at $x = 0$, but holds a mass M at $x = L$. This system has normal modes, but as we shall see they are no longer harmonically related. The pressure to the right of the mass is assumed constant. The boundary conditions are

$$u_1 = 0 \quad \text{at } x = 0 \qquad M(\partial u_1 / \partial t) = A p_1 \quad \text{at } x = L. \tag{4.26}$$

Figure 4.4 A pipe closed at $x = 0$ and containing a mass M at $x = L$.

We can express the latter in terms of u_1 by differentiating it with respect to t and using equations (4.6) and (4.8).

$$M(\partial^2 u_1 / \partial t^2) = -A c^2 \rho_0 (\partial u_1 / \partial x) \qquad \text{at } x = L. \tag{4.27}$$

The separable solution satisfying the first condition is, as before

$$u_1 = C \cos \omega t \sin(\omega x/c). \tag{4.28}$$

If we substitute this expression into the second condition, we obtain

$$-M\omega^2 \sin(\omega L/c) = -Ac\rho_0 \omega \cos(\omega L/c)$$

or

$$\cot(\omega L/c) = M\omega/A\rho_0 c. \quad (4.29)$$

This equation determines the frequencies ω of the normal modes. Notice that the expression on the right is the quantity μ introduced in equation (4.20). Equation (4.29) has two limiting cases. When $M=0$ it gives $\cos(\omega L/c)=0$ which is equation (4.10) for an *open* pipe. When $M \to \infty$, it gives $\sin(\omega L/c)=0$, which is easily shown to be the appropriate modal equation for a pipe *closed* at both ends.

For general values of M, (4.29) is a *transcendental* equation; it involves ω both algebraically and inside a trigonometric function. It cannot be solved exactly analytically. Numerical solution is necessary for detailed results, but approximations can be made as follows. Introduce first an angle θ

$$\theta = \omega L/c_0. \quad (4.30)$$

Then our equation can be written

$$\cot \theta = v\theta \quad (4.31)$$

where

$$v = M/AL\rho_0. \quad (4.32)$$

'Small' and 'large' mass in this context is to be measured by the size of v relative to unity. Note that $AL\rho_0$ is just the mass of gas in the pipe.

Figure 4.5 shows a graph of $\cot \theta$ for $0 \leqslant \theta \leqslant 4\pi$. Additionally the line $v\theta$ with slope v is drawn. The intersections of these occur at the roots of equation (4.31). It is obvious that there is an infinite number of these, and so an infinite number of normal modes. Suppose that v is small. Then the graph shows that θ_0 is slightly less than $\pi/2$. So let

$$\theta_0 = \tfrac{1}{2}\pi - \delta \quad (4.33)$$

where δ is small. Then $\cot \theta_0 = \tan \delta \simeq \delta$, and so (4.31) is approximately $\delta \simeq v(\tfrac{1}{2}\pi - \delta)$ or $\delta \simeq \tfrac{1}{2}\pi v/(1+v) \simeq \tfrac{1}{2}\pi v$. Thus

$$\theta_0 \simeq \tfrac{1}{2}\pi(1-v). \quad (4.34)$$

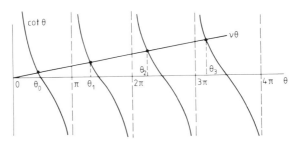

Figure 4.5 Graphical solution of equation (4.30).

Sound Waves

The fundamental frequency ω_0 is, from (4.30)

$$\omega_0 \simeq \tfrac{1}{2}\pi c_0 L^{-1}(1-v) \tag{4.35}$$

and is reduced by an amount proportional to the mass M. For $A = 10^{-3}$ m, $L = 1$ m, $\rho_0 \simeq 1$ kg m^{-3}, $M = 5 \times 10^{-5}$ kg, we have $v = 0.05$ and so ω_0 is reduced by 5%.

It is clear from figure 4.5 that the low-order modes $n = 1, 2, 3, \ldots$ have values of θ that differ more and more from odd multiples of $\tfrac{1}{2}\pi$. These usually have to be obtained numerically. However, for high enough mode numbers, the values of θ approach multiples of π from above. Physically this implies that the mass M, however small, will behave like a rigid wall at sufficiently high frequency, and the modes of a closed pipe will be approached. (The inertial effect of M is proportional to $M\omega^2$, and no matter how small is M this can become dominant if ω is large enough.) To estimate these θ values, let

$$\theta_n = n\pi + \delta \tag{4.36}$$

where δ is small.

So $\cot \theta = \cot \delta \simeq \delta^{-1}$, and (4.31) becomes $\delta^{-1} \simeq v(n\pi + \delta)$. Hence

$$\delta \simeq (n\pi v)^{-1}. \tag{4.37}$$

So

$$\theta_n \simeq n\pi + (n\pi v)^{-1}$$

and

$$\omega_n \simeq c_0 n\pi L^{-1}[1 + (n^2\pi^2 v)^{-1}]. \tag{4.38}$$

So the frequencies of the high-order modes differ from those of a closed pipe by an amount proportional to n^{-2}. This approximation will be valid, say $\delta < 0.1$ radian, provided

$$(n\pi v)^{-1} < 0.1 \quad \text{or} \quad n \gtrsim 3/v. \tag{4.39}$$

So if $v = 0.05$, the modes above the 60th are described by (4.38), but if M is increased in our example to 10^{-3} kg then $v = 1$ and modes above the third are so described and, for all but the first two or three modes, the system behaves like a closed pipe.

This concludes our discussion of linear sound waves.

Waves on Water 5

5.1 Introduction

The most familiar waves are those on the surface of water. Ripples on puddles of wavelength 10^{-3} m and ocean waves of length greater than 100 m are extreme examples. The pattern of waves behind a ship is an intriguing problem that will be outlined in this chapter. The linearised theory of water waves is, rather ironically, often *not* described by our familiar wave equation. Tidal waves, tidal bores and breakers indicate that the non-linear theory is of importance, but this, a subject of current active research, is beyond the scope of this book.

In order to derive adequately the equations for water waves it is necessary to consider fluid mechanics in detail, and there is no scope for this here. We shall consider two limiting cases; waves on water that is respectively shallow and deep compared with the wavelength. In the former case, a full set of equations is readily derived (§5.2) but in the latter we shall have recourse to dimensional analysis. It will turn out that the deep water ship wave pattern can be analysed using only these dimensional results.

An aspect of water waves that passes unnoticed because of its familiarity is that they are usually *wavy*. A stone dropped into water produces a pattern of ripples regardless of its shape or size or manner of impact with the water. In contrast, oscillatory solutions of the wave equation in Chapters 3 and 4 appeared only when special boundary or initial conditions were imposed. There is a very important reason for this which will appear later.

From the outset we shall make some approximations. Variations in the *density* of the water will be neglected. Of course, sound waves, which incorporate density changes, can propagate in water, but we shall not consider them. We shall also neglect *surface tension*, and *bottom friction*. Fluid mechanics texts discuss the validity of such approximations.

5.2 Waves on Shallow Water

Consider a uniform rectangular channel of width b with vertical side walls. It is filled with water of fixed density ρ, to some reference height h_0 above the horizontal bottom. Disturbances are represented as variations $h(x, t)$ in the height, where x is distance along the channel. A reasonable first approximation to rivers and canals is provided by this model, if the coordinate system is assumed to travel with the mean speed of the flow. The analysis of the problem parallels that of sound waves in a pipe in Chapter 4. We consider (figure 5.1) the water between two fixed vertical planes at x and $x + \Delta x$. These define a *reference volume*. We shall assume that the velocity u of water in the x direction is independent of depth, and so columns of water move as a whole. Vertical motions of water are neglected. Our two variables are h and u, and the two equations needed to determine them follow from conservation of mass and momentum.

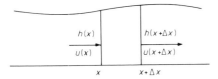

Figure 5.1 The reference volume of water.

5.2.1 *Conservation of Mass*

At time t, the mass of water in the reference volume is $\rho bh \, \Delta x$. At time $t + \Delta t$ this will increase by the amount $\rho b (\partial h/\partial t) \, \Delta t \, \Delta x$ if Δt is small enough. This increase is met by a net inflow across the planes x and $x + \Delta x$. That *into* x in time Δt is $\rho buh \, \Delta t$ evaluated at x. That *out of* $x + \Delta x$ is $\rho buh \, \Delta t$ evaluated at $x + \Delta x$. The net inflow is the difference $-\rho b (\partial (uh)/\partial x) \, \Delta t \, \Delta x$, if Δx is small enough. Equating these expressions we obtain the equation of *mass conservation* or *continuity*

$$\partial h/\partial t = -\partial (hu)/\partial x. \qquad (5.1)$$

The analogy with the sound wave analysis of Chapter 4 is obvious. The height h corresponds to the density ρ in sound theory.

5.2.2 *Conservation of Momentum*

The momentum of the water in the reference volume at time t is $\rho bhu \, \Delta x$. Calculation of the rate of change of this with time and its compensation by inflow of momentum through the planes x and $x + \Delta x$ parallels precisely that for mass conservation. However, the net force of pressure across x and $x + \Delta x$ affects the change in momentum. Let this force be $F(x, t)$. Then it is easily shown that

$$\partial (hu)/\partial t = -\partial (hu^2)/\partial x - (\rho b)^{-1} \, \partial F/\partial x. \qquad (5.2)$$

To calculate F, we assume that the law of *hydrostatics* is obeyed: that the pressure at depth ζ below the surface is

$$p = p_0 + \rho g \zeta \qquad (5.3)$$

where p_0 is the atmospheric pressure and g is the acceleration due to gravity. This assumption is related to the neglect of vertical motions (and hence accelerations) already mentioned. p_0 acts over the whole surface of the reference volume and has no net effect. F is the resultant of the excess pressure $\rho g\zeta$ at all depths from 0 to h

$$F = \int_0^h \rho g\zeta b d\zeta = \tfrac{1}{2}\rho g b h^2. \tag{5.4}$$

Then (5.2) becomes

$$\partial(hu)/\partial t = -\partial(hu^2)/\partial x - gh\,\partial h/\partial x. \tag{5.5}$$

Again the parallel with the corresponding equation for sound is clear. Equation (5.5) is the expression of *conservation of momentum* or simply the *equation of motion*.

We can use (5.1) to simplify (5.5), and obtain finally the pair of equations

$$\partial h/\partial t + \partial(hu)/\partial x = 0 \qquad \partial u/\partial t + u(\partial u/\partial x) + g(\partial h/\partial x) = 0. \tag{5.6}$$

We now concentrate on the linearised approximation to these.

5.3 Linearised Waves on Shallow Water

We write

$$h = h_0 + \eta \tag{5.7}$$

where $|\eta| \ll h_0$. Substituting into equations (5.6) and neglecting squares and products of η and u

$$\partial \eta/\partial t + h_0(\partial u/\partial x) = 0 \qquad \partial u/\partial t + g(\partial \eta/\partial x) = 0. \tag{5.8}$$

If u is eliminated we get

$$\partial^2 \eta/\partial t^2 = c^2(\partial^2 \eta/\partial x^2) \tag{5.9}$$

where

$$c = (gh_0)^{1/2}. \tag{5.10}$$

u therefore obeys the familiar wave equation. The wave speed c varies from 1 to 10 m s^{-1} as h_0 varies from 0.1 to 10 m. Such values seem not unreasonable for water waves.

It will not be necessary to consider new solutions to (5.9). Necessary ones can simply be taken from previous chapters. For instance, consider a tank of length L. The velocity u has to be zero at $x = 0$ and $x = L$. Sinusoidal normal modes exist in which an odd number of half wavelengths fit into the length L. Their frequencies are given by

$$f = \pi c L^{-1}(2n+1) \tag{5.11}$$

where n is an integer. If $h_0 = 0.1$ m, then $c = 1$ m s^{-1}. For a tank of length 10 m the fundamental frequency is about 1/3 Hz (an oscillation every three seconds). Our theory is valid only if h_0 is much less than the wavelength $\lambda = c/f$. In this example $\lambda > h_0$ up to the 15th harmonic.

5.4 Waves on Deep Water

The fluid dynamics necessary for the description of disturbances on deep water is beyond the scope of this book. Water motions turn out to be confined substantially to a layer adjacent to the surface about one wavelength in depth, and both horizontal and vertical motions are important in this layer. However, the key feature of such waves can be derived using dimensional analysis only. Returning briefly to the case of shallow water, suppose we are asked to estimate the wave speed without performing any mechanical analysis. How can a *velocity* be constructed from the physical parameters of the problem? These are three in number: g, h_0 and ρ, having respective physical dimensions, LT^{-2}, L and ML^{-3}, where L, T, M denote length, time and mass. The only quantity having the dimensions LT^{-1} of a velocity that can be constructed from these is $(gh_0)^{1/2}$. It is therefore plausible that the relevant waves have a speed proportional to this. That the constant of proportionality is unity (5.10) follows only from mechanical analysis.

What is the corresponding argument for deep water? It is conceivable that the same formula should apply. On the deep ocean, however, h_0 is of the order 10 km, yielding $(gh_0)^{1/2} \sim 300$ m s^{-1}. Obviously, ocean waves have speeds vastly lower than this. If h_0 is of this order, it should not enter the problem at all. There remain only g and ρ as physical parameters, and it is impossible to construct a quantity with dimensions of velocity from these alone. We conclude that *there is no unique speed for waves on deep water*. This is a result of fundamental importance, and its consequences give rise to most of the intriguing properties of water waves.

Ignoring for the moment the wave speed, let us imagine a sinusoidal wave with frequency ω and wavenumber k. Can we construct a relation between ω and k that involves only g and ρ? Since ρ involves *mass* it is clear that the only such relation is

$$\omega^2 = Kgk \tag{5.12}$$

where K is a dimensionless constant. The complete linearised dynamical analysis confirms this, and shows that $K = 1$, so that

$$\omega^2 = gk. \tag{5.13}$$

The wave speed $c = \omega/k$ is given by

$$c = (g\lambda/2\pi)^{1/2} \tag{5.14}$$

where $\lambda = 2\pi/k$ is the wavelength. *The speed depends on the wavelength.* Longer waves travel faster. For instance, if $\lambda = 1$ m, $c = 1.3$ m s^{-1} and if $\lambda = 100$ m, $c = 13$ m s^{-1}. The corresponding frequencies are 1.3 and 0.13 Hz, and such results seem not unreasonable.

Before considering the consequences of (5.12) and (5.13), we state the corresponding result for waves in water of arbitrary depth. This is quite beyond the scope of dimensional analysis. It is

$$\omega^2 = gk \tanh(kh_0). \tag{5.15}$$

On *shallow* water $kh_0 \ll 1$, and since $\tanh x \simeq x$ if x is small, (5.15) becomes $\omega^2 = gh_0 k^2$, or $c = \omega/k = (gh_0)^{1/2}$. For *deep* water, kh_0 is large, $\tanh kh_0 \simeq 1$, and (5.13) is recovered. For intermediate depths, if $c_0 = (gh_0)^{1/2}$ then we can write

$$c = c_0[\tanh(2\pi h_0/\lambda)/(2\pi h_0/\lambda)]^{1/2}. \tag{5.16}$$

c/c_0 is plotted against λ/h_0 in figure 5.2. The shallow water result $c = c_0$ holds to within 3% if λ/h_0 exceeds about 15.

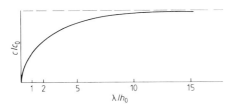

Figure 5.2 Speed/wavelength relation for water waves.

We return to (5.13). A localised disturbance can be regarded as a superposition of sinusoidal components with a range of wavelengths. On deep water, each of these will travel at a different speed. The disturbance will therefore *disperse* as it propagates. Component waves will be found in different places. This is the reason for the pattern of ripples caused by a stone dropped into a pond. The longer waves travel faster, and appear on the outside of the pattern, followed by the shorter ones. In contrast, a disturbance on shallow water will spread out as a single front, without accompanying ripples. With a little care, this can be seen if a saucer is immersed gently into shallow water in a large flat trough.

A wave system in which speed depends on wavelength is called *dispersive*, and the relation between ω and k is the corresponding *dispersion equation*. The crucial property of such systems emerges when we consider the superposition of two or more waves with different frequencies. Consider therefore a disturbance consisting of two sinusoids of equal amplitude

$$d = a[\cos(\omega_1 t - k_1 x) + \cos(\omega_2 t - k_2 x)] \tag{5.17}$$

or equivalently

$$d = 2a \cos(\Omega t - Kx) \cos(\omega t - kx) \tag{5.18}$$

where $\Omega = (\omega_2 - \omega_1)/2$, $K = (k_2 - k_1)/2$, $\omega = (\omega_2 + \omega_1)/2$ and $k = (k_2 + k_1)/2$.

Suppose that k_1 and k_2 are nearly equal, so that $K \ll k$. At a fixed value of t, d varies with x as indicated in figure 5.3. It consists of a *carrier wave* of wavelength $2\pi/k$, slowly modulated by the first term in (5.18), which alters the amplitude of the carrier from 0 to $2a$ over a distance π/K. If the system is *not* dispersive then ω_1/k_1 and ω_2/k_2 are each equal to the constant wave speed c, as are ω/k and Ω/K. The whole pattern in the figure will move to the right with speed c. If it *is* dispersive though, ω_1/k_1 and ω_2/k_2 will differ slightly. If the dispersion equation gives ω as a function of k, and if $k_2 - k_1$ is small, the speed of the carrier wave is

$$c = \omega/k \simeq \omega_1/k_1 \simeq \omega_2/k_2 \tag{5.19}$$

but the modulation has the speed

$$u = \Omega/K = (\omega_2 - \omega_1)/(k_2 - k_1) \to d\omega/dk. \tag{5.20}$$

On deep water (5.12) yields

$$u = d\omega/dk = \tfrac{1}{2}(g/k)^{1/2} = \tfrac{1}{2}c \tag{5.21}$$

so the modulations move at *half* the speed of the carrier.

A *wave packet* is a localised modulation of a carrier wave. It is similar to one cycle of the modulations in figure 5.3, and can be shown to be composed of *many* (as opposed to two) sinusoids of nearly equal wavelength. The foregoing remarks apply here as well. The underlying oscillations of the carrier travel at speed $c = \omega/k$, the modulations at speed $u = d\omega/dk$, where k is the average wavenumber of the component sinusoids. c is called the *phase velocity* and u the *group velocity*. If $u = \tfrac{1}{2}c$, the oscillations will appear at one end of the packet, travel through it, and disappear at the other end.

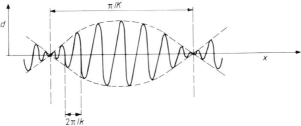

Figure 5.3 Superposition of two sine waves with nearly equal wavelengths.

On water of depth h, the reader can show from (5.15) that

$$u/c = \tfrac{1}{2}[1 + 2kh/(\sinh 2kh)]. \tag{5.22}$$

On deep water $kh \to \infty$ and $u/c \to \tfrac{1}{2}$, while on shallow water $kh \to 0$ and $u/c \to 1$. This is expected, since shallow water waves are not dispersive. Generally, u/c can take any value, and can even be negative. In that case, the carrier and its modulations would move in opposite directions.

Granted these statements, it is not unreasonable to suppose that the *energy* associated with a dispersive wave system will travel at the group velocity. This is indeed the case, though any formal demonstration of its truth is beyond the scope of this book.

We shall conclude this chapter with an example of the power of the concept of group velocity: the analysis of the ship wave pattern. Although this is not one dimensional, the discussion so far is adequate for the calculation of its shape.

5.5 Ship Waves

Consider a moving point disturbance (a ship!) on deep water. Remarkably, the shape of the pattern depends solely on the fact that $u = \frac{1}{2}c$ for all wavelengths. Recalling the dimensional analysis of §5.3 that led to the dispersion equation (5.11), we see that the pattern can be derived without using any fluid mechanics at all. The constant K in (5.11) does not affect the value of u/c.

A ship generates waves at each point of its path. Those observed in the steady pattern are the ones that keep precisely in step with the ship. In figure 5.4, a wavelet travels at speed c away from its origin at A on the ship's path. The ship has speed V to the left, and that part of the wavefront that will keep in step with the ship is at C, and is such that the component of V in the direction AC is equal to c

$$c = V \cos \theta. \tag{5.23}$$

If the waves were non-dispersive, c would be constant, and if $V > c$ a unique direction θ is determined. Waves emitted at successive instants that keep in step with the ship would have as their envelope a pair of straight lines emerging from the ship at an angle $\alpha = \sin^{-1} c/V$, as shown in figure 5.5. A supersonic projectile $(c > V)$ in air is an example of this, since sound waves are not dispersive. The resulting disturbance (a *cone* of semi-angle α) is in fact a *shock wave*.

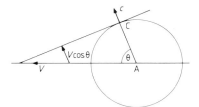

Figure 5.4 Wavelet that keeps in step with ship.

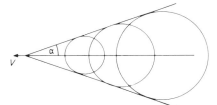

Figure 5.5 Non-dispersive waves from a moving source.

For dispersive waves, however, the behaviour is quite different. In time t the ship moves (figure 5.6) a distance Vt from A to B. Waves are generated at A with various speeds and directions, and several of those that keep in step with the ship are illustrated. Larger, faster waves reach C_1 while shorter, slower ones reach C_4. From (5.23) it follows that points C_1 to C_4 lie on a circle with diameter AB.

Although the individual oscillations will be at C_1 to C_4, the corresponding visible water disturbances travel at the group velocity $\frac{1}{2}c$, and so will be at locations D_1 to D_4 in figure 5.7. These points also lie on a circle, but with diameter $\frac{1}{2}Vt$. The visible parts of the crests are each at right angles to the corresponding lines AD and the pattern is composed of all such crests emitted at successive past positions of the ship. The various circles on which the points D lie at different past positions clearly have a common tangent, the line BE in figure 5.8. The visible pattern lies therefore within a wedge of angle 2α, where α is given by

$$\sin \alpha = \text{EF}/\text{BF} = \tfrac{1}{4}Vt/\tfrac{3}{4}Vt = \tfrac{1}{3} \qquad \text{or} \qquad \alpha \simeq 19.5° \tag{5.24}$$

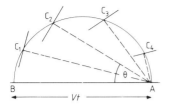
Figure 5.6 Ship waves: phase fronts.

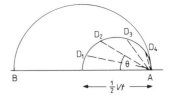
Figure 5.7 Ship waves: visible disturbances moving at group velocity.

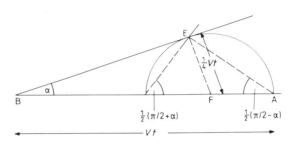
Figure 5.8 The Kelvin wedge.

and this is independent of V. The region of disturbance is known as the *Kelvin wedge*. This remarkable result has been deduced solely from (5.23) and the fact that $u/c = \frac{1}{2}$.

From (5.14) and (5.23) the wavelength visible in direction θ is

$$\lambda = (2\pi V^2/g) \cos^2 \theta \qquad (5.25)$$

and so those waves travelling parallel to the ship ($\theta = 0$) have the longest wavelength $\lambda_m = 2\pi V^2/g$, which is 60 m if $V = 10 \text{ m s}^{-1}$. On the edge of the pattern, at E in figure 5.8, we have $\theta = \frac{1}{2}(\frac{1}{2}\pi - \alpha) = \cos^{-1}\sqrt{2/3}$ and $\lambda = 2\lambda_m/3$. These limiting wave crests have an angle $\frac{1}{2}(\frac{1}{2}\pi + \alpha) = \sin^{-1}(1/\sqrt{3}) = 55°$, with the direction of travel. They are frequently the most readily visible part of the pattern, since the interior of the wedge is confused by the presence of two sets of waves.

The detailed appearance of the pattern depends on the range of wavelengths that is generated by the ship, which depends in turn on the ratio of its size and λ_m. The shape can be calculated easily, but the amplitude distribution, which determines the appearance, needs much more intricate analysis.

Take the origin of coordinates at the current position of the ship. Waves launched at $(X, 0)$ at angle θ will appear at (x, y) where

$$x = X(1 - \tfrac{1}{2}\cos^2 \theta) \qquad y = X\tfrac{1}{2}\cos\theta \sin\theta. \qquad (5.26)$$

This is derived from the construction in figure 5.7 and illustrated in figure 5.9. The pattern is the envelope of corresponding wavefronts as θ varies. Each of these is locally at 90° to AD in figure 5.9, and so has gradient

$$dy/dx = \cot \theta. \tag{5.27}$$

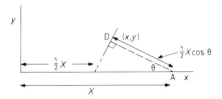

Figure 5.9 Wave pattern coordinates.

Different parts of the envelope will have been launched at different values of X. To find its equation, we must use (5.26) in (5.27) to relate X to θ. After some algebra, we get

$$X^{-1}(dX/d\theta) = -\tan \theta \tag{5.28}$$

which is a separable ordinary differential equation integrating to

$$X = A \cos \theta \tag{5.29}$$

where A is a constant. The pattern is then given in parametric form by (5.26) and (5.29)

$$x = A \cos \theta (1 - \tfrac{1}{2} \cos^2 \theta) \qquad y = A \tfrac{1}{2} \cos^2 \theta \sin \theta. \tag{5.30}$$

The result is the cusped curve in figure 5.10, and the superposition of several of these with different values of A (which are all of similar shape) gives the complete pattern. The details of this depend on the ship's size and are intricate to evaluate, but nevertheless the pattern in figure 5.10 is substantially as often observed.

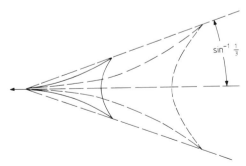

Figure 5.10 The ship wave pattern.

This concludes our discussion of water waves, which has served to introduce the vital concepts of dispersion and group velocity. It also concludes this introductory monograph on the mathematics of oscillations and waves.

Index

Adiabatic law for gas, 38, 39
Air, 37
Amplitude of oscillation, 3
   of wave, 33

Bandwidth of oscillator, 15, 16
Beats, 21
Binomial theorem, 7
Bottom friction, 46
Boundary conditions, 22, 24, 26, 31, 34, 40, 48

Carrier wave, 50
Complex amplitude, 3
Complex exponential function, 3
Complimentary function, 13
Conservation of energy, 6, 38, 43
   of mass, 38, 47
   of momentum, 38, 47
Constants of integration, 2, 3, 12, 18
Continuity equation, 38, 47
Continuous string, 25
Coupled systems, 17, 19, 20

Damping, linear, 8, 10, 12–14
Decibel, 40
Degrees of freedom, 18
   infinite number of, 23
Density of gas, 37
Differential equation, ordinary, 2
   partial, 26–31
Dimensional analysis, 46, 49
Dispersion equation, 50
Dispersive waves, 40, 50, 51
Drum, 35

Energy, 27, 35, 36
   conservation of, 6, 38, 43
   kinetic, 6, 35, 36
   potential, 6, 35, 36

Equations, algebraic, 23
   coupled, 18, 23
   linear, 8, 22, 30
   non-linear, 2, 5, 30, 39, 48
   transcendental, 44
   uncoupled, 18
Equations of motion, 2, 5, 6, 8, 18, 22, 27, 29, 38, 48

Force, 2
   periodic, 12
Fourier analysis, 35, 40
Frequency, 4
   angular, 4
   forcing, 13
   natural, 21
Friction, 7, 9
Functions of integration, 28
Fundamental, 34, 40, 45

Gas, 37
Gravity, 1, 18, 19, 29
Group velocity, 51–4

Harmonics, 34
Hooke's law, 17, 22, 30
Hydrostatics, 47

Impedance, mechanical, 13
Impulse, 11, 12
Inertia, 13, 14, 29
Initial conditions, 2, 3, 13, 18, 20, 29, 35
Intensity of sound waves, 41

Kelvin wedge, 53

Linear equation, 8, 22, 30
Linearisation of equations, 2, 5, 17, 39, 40, 48
Linearity, 2, 8, 29

55

Mass, conservation of, 38, 47
Mode, see Normal modes
Modulation, 21, 51
Molecules, 37
Momentum, 37
  conservation of, 38, 47

Newton's law of motion, 1, 28
Normal modes of oscillation, 17, 19–26
  of sound waves in pipe, 40–5
  of stretched string, 31, 34–6
  of water in a tank, 48

Oscillations, 1, 11–17
  damped, 7, 11, 12
  transverse, 22
Oscillator, 8, 13, 15, 16
  coupled, 16
  critically dampled, 9, 11
  forced, 12
  overdamped, 9, 10
  underdamped, 9, 10

Partial derivative, 25
Particular integral, 13
Pendulum, 1, 5, 17–19
  coupled, 17, 22, 34
Period of oscillation, 4
Phase, 5
  velocity, 51
Pipe, 39–41
Power consumption of oscillator, 15
Pressure in gas, 37, 40–2
  gradient, 37
Pulse, 31

Quality factor of oscillator, 15, 16

Recurrence relation, 23
Reference volume of gas, 38
  water, 47
Resonance, 12, 15

Separable solutions of wave equations, 32, 40
Ship waves, 46, 51–4
Shock wave, 52
Simple harmonic motion (SHM), 4, 5, 10, 18, 23
Sound, 35, 37
  speed of, 39
  Specific heat ratio of gas, 39
Spring, 17–19
Steady state response, 13
Stiffness, 13
Stretched string, 22, 26–9
Surface tension, 46

Taylor's theorem, 25
Tension, 1, 17, 18, 22, 29, 30
Transient, 13, 16

Waves, 23, 26, 27, 31
  on deep water, 49–51
  equation, 28–31, 35, 39, 46, 48
  length, 33
  motion, 25
  number, 33
  packet, 51
  reflection, sound, 41
  reflection, stretched string, 31
  on shallow water, 46–8
  sinusoidal, 31, 33, 36, 41
  sound, 37–45
  standing, 31, 33
  velocity of, 25–7, 30, 33, 36, 39
  water, 27, 46–54